ACS SYMPOSIUM SERIES 275

Nutritional Bioavailability of Calcium

Constance Kies, EDITOR
University of Nebraska—Lincoln

Based on a symposium sponsored by
the Division of Agricultural and Food Chemistry
at the 187th Meeting
of the American Chemical Society,
St. Louis, Missouri, April 8–13, 1984

American Chemical Society, Washington, D.C. 1985

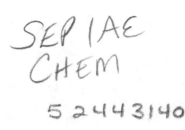

SEP IAE
CHEM

5 2443140

Library of Congress Cataloging in Publication Data

Nutritional bioavailability of calcium.
(ACS symposium series, ISSN 0097–6156; 275)

"Based on a symposium sponsored by the Division of
Agricultural and Food Chemistry at the 187th Meeting
of the American Chemical Society, St. Louis, Missouri,
April 8–13, 1984."

Includes bibliographies and indexes.

1. Calcium—Absorption and adsorption—
Congresses. 2. Calcium—Metabolism—Congresses.
3. Intestinal absorption—Congresses.

I. Kies, Constance, 1934–   . II. American
Chemical Society. Meeting (187th: 1984: St. Louis,
Mo.) III. American Chemical Society. Division of
Agricultural and Food Chemistry. IV. Series.

[DNLM: 1. Biological Availability—congresses.
2. Calcium—metabolism—congresses. 3. Nutrition—
congresses. QV 276 N976 1984]

QP535.C2N88   1985   599'.0133   85–3931
ISBN 0–8412–0907–3

FOREWORD

The ACS SYMPOSIUM SERIES was founded in 1974 to provide a medium for publishing symposia quickly in book form. The format of the Series parallels that of the continuing ADVANCES IN CHEMISTRY SERIES except that, in order to save time, the papers are not typeset but are reproduced as they are submitted by the authors in camera-ready form. Papers are reviewed under the supervision of the Editors with the assistance of the Series Advisory Board and are selected to maintain the integrity of the symposia; however, verbatim reproductions of previously published papers are not accepted. Both reviews and reports of research are acceptable, because symposia may embrace both types of presentation.

CONTENTS

PREFACE

High incidence of osteoporosis among postmenopausal women has created an intensified interest in adequacy of calcium nutritional status among Americans of all ages and sexes. To eat an assumed adequate amount of a nutrient does not necessarily guarantee dietary sufficiency of that nutrient. The degree to which a nutrient is absorbed from the intestinal tract, its efficiency of utilization within the body, and the processes governing its excretion are all contributing factors.

The importance of dietary or endogenously synthesized vitamin D has long been recognized as a primary factor influencing the bioavailability of calcium. Some of the most exciting biochemical–nutritional research in recent years has been devoted to determining the mechanisms involved in vitamin D–calcium interactions. This research has been well reviewed in other publications. The objective of the symposium upon which this book is based was to review some of the other lesser-known dietary factors that appear to have an impact on the bioavailability of calcium.

For helping plan the symposium and for handling many of the details involved in the preparation of this book, I would like to acknowledge the assistance and work of my secretary, Mrs. Donna Hahn.

CONSTANCE KIES
University of Nebraska—Lincoln

December 17, 1984

Dietary Calcium Exchangeability and Bioavailability
Evaluation and Potential Uses of an In Vitro Digestion Procedure

E. M. WIEN and RUTH SCHWARTZ

Division of Nutritional Sciences, Cornell University, Ithaca, NY 14853

Use of the in vitro digestion procedure for estimating
dietary calcium exchange with an extrinsic isotope
should facilitate in vivo absorption studies using the
"extrinsic tag" technique. Only a peptic digestion
stage is required for exchangeability measurement.
Attempts to extend use of the the procedure to measure
parameters of bioavailability by including a pancreatic
digestion stage were partly successful. In vitro digestion permits study of the chemistry of food calcium
under standardized digestion conditions. Investigations discussed include the effects of varying pH,
bile salts, enzymes and food substrates on calcium
solubility; and post-digestion fractionation of calcium
complexes. Before bioavailability per se is estimated
in vitro, more direct comparisons between in vivo and
in vitro measurements are needed.

The concept of bioavailability was developed to explain the
difference between the total amount of mineral in a food and the
amount which was used by the individual consuming the food. Over the
past sixty years or more, there have been numerous studies related to
dietary calcium requirements and bioavailability (1,2). As a result,
much is known about non-calcium food components which influence the
absorption and utilization of dietary calcium under experimental
conditions. What now is lacking is a detailed knowledge of how these
factors interact with calcium under normal conditions of ingestion in
meals.

We have developed an in vitro digestion procedure, not as a
substitute for in vivo studies, but as a useful adjunct. Our initial
objective was to develop an in vitro procedure for measuring exchangeability, the fraction of the food mineral which exchanges with an
extrinsic isotope tracer added to the food. This was expected to
facilitate the measurement of food mineral absorption in humans by the
extrinsic tag method. Secondary objectives were to determine if in
vitro mineral solubility could be used to estimate potential

bioavailability and to explore ways of using the in vitro procedure to study the interactions of gastrointestinal secretions with dietary mineral and other dietary components which determine mineral solubility. A number of in vitro digestion procedures have been developed for estimating iron bioavailability (e.g. 3-5) and nonheme iron exchangeability (6). The procedure discussed in this paper evolved from one of them (4) for studying other minerals. Most of the work in our laboratory has been done with calcium.

In this report our original procedure (7) is described. Then the published in vivo - in vitro comparison experiments (8) are briefly summarized and subsequent investigations on in vitro digestion conditions and methods for fractionation of calcium from digests are reported.

In Vitro Digestion Procedure

The in vitro digestion procedure is outlined in Figure 1. The pepsin, pancreatin and two conjugated bile salts mixtures - porcine bile extract (BE) and bovine "sodium taurocholate" (TC) - were obtained from Sigma Chemical Co. (St. Louis, MO). The commercial preparations were analyzed for mineral content and decontaminated if necessary before use in the procedure. Usually BE was used in experiments discussed in this paper. In some experiments the digestion was stopped after peptic digestion. Results of analyses made after the peptic digestion only will be referred to as "after the P stage"; results of analyses made after the complete peptic + pancreatic digestion will be referred to as "after complete digestion" or "after PPa digestion."

Exchangeability is calculated as the ratio of specific activities (dpm [45]Ca/ug Ca) of the mixture and supernatant, when the extrinsic isotope was added in ionizable form (Figure 1). Exchangeability values are expressed as decimal fractions.

The in vitro estimate of potential availability was defined, somewhat arbitrarily, as calcium solubility (18,000 x g supernatant) after complete digestion. Potentially available calcium was expressed as a percentage of the total food calcium (Figure 1). With the exception of a low in vitro calcium solubility value for whole milk, our earlier data compared reasonably well with calcium bioavailability information in the literature (7).

Comparison of In Vivo and In Vitro Measurements

The in vitro procedure was tested in "critical" experiments designed to make direct comparisons of in vivo and in vitro estimates of exchangeability and potential bioavailability and to test the use of in vitro exchangeability values in in vivo experiments. (8). Three foods which were expected to show different levels of calcium solubility and exchangeability, collards, soybeans and spinach, were intrinsically labeled with [47]Ca in nutrient solution culture. They were used together with [47]Ca as an extrinsic label in both in vitro and in vivo experiments.

The results (8) showed the expected variation in exchangeable Ca among the foods; exchangeability was not complete for soy or spinach. However, in vivo and in vitro exchangeability values were nearly identical for each food. The in vitro exchangeability values after

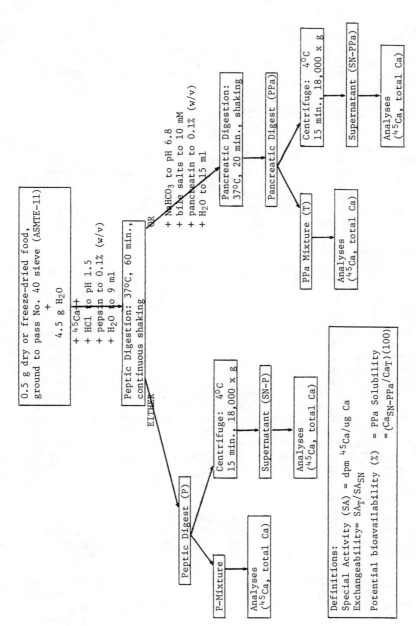

Figure 1. In vitro digestion procedure. (Reproduced with permission from Ref. 7. Copyright 1982 J. Nutr., American Institute of Nutrition.)

PPa digestion were used to correct in vivo extrinsic tag absorption in order to estimate intrinsic food calcium absorption. The corrected extrinsic tag absorption agreed well with the intrinsic tag absorption in rats for all three foods.

When the estimates of bioavailability were compared (8), in vivo absorption was higher than in vitro solubility for two of the foods: We had expected absorption to be less than solubility due to physiological factors (1,9). Thus, this surprising result led to the reexamination of in vitro digestion conditions which is reported in this paper.

Review of In Vitro Digestion Conditions

Experiments were conducted to determine if varying the conditions in the in vitro digestion procedure would affect post-digestion calcium solubility and in some cases, exchangeability. This was done with two purposes: to test the use of the in vitro digestion procedure for studying factors which might influence calcium bioavailability and to use the results to modify the standard procedure.

Weights and volumes were doubled from the original procedure (Figure 1) to have more material for analysis.

Peptic Digestion

Only initial pH of the peptic digestion was studied. The pH was set either at 1.5, near the pH optimum for pepsin activity (10) or 2.0, to keep the pH closer to the range observed during human gastric digestion (11). The results in Table I show the progressive changes in calcium "exchangeability" and solubility from the initial slurry through peptic and pancreatic digestion for two cow's milk products and four soy products. Varying the peptic pH between 1.5 and 2.0 had little effect on exchangeability and solubility at either the peptic or pancreatic stage. There was little relationship between the initial slurry and post-digestion values. Exchangeability was determined at the peptic stage, but was incomplete for three of the foods. Little change occurred during pancreatic digestion. Solubility was maximum after peptic digestion and decreased during pancreatic digestion for four of the foods. Since the exchangeability did not change during the pancreatic digestion while solubility decreased, the food calcium and extrinsic tag must have precipitated from solution at the same rate during pancreatic digestion.

Pancreatic Digestion

The pancreatic digestion conditions studied included pH, the method of pH control, and bile salts mixture and concentration. In addition, experiments were run to determine if mineral solubility was affected by enzymatic activity, or only by pH-induced solubility changes.

pH and pH Control. In the original procedure (Figure 1), the pH was adjusted to pH 6.8 with freshly prepared $NaHCO_3$. The pH generally rose during pancreatic digestion but the magnitude varied with different foods and pH adjustment techniques. The data in Figure 2 are from a number of experiments in which the pH at the end of pancreatic digestion varied from about 6.2 to 7.2. Although calcium

Table I. In Vitro Calcium Exchangeablility and Solubility in Milk and Soy Products as Affected by Stage of Digestion and Initial Peptic pH

Substrate	mg Ca per flask[1]	pH at Stage[2]				Exchangeability, SA ratio[3,4]			Solubility, %[3,5]		
		I	Po	P	PPa	I	P	PPa	I	P	PPa
Skim Milk	12.40	6.7	1.5	1.8	6.7	0.90±.04	1.00±.02	0.97±.01	64.7±2.3	95.4±1.8	93.7±1.8
	12.40	6.7	2.0	2.4	6.6	0.90±.04	0.96±.02	1.02±.03	64.7±2.3	93.3±5.0	91.7±1.7
Whole Milk	9.72	6.7	1.5	1.8	6.5	0.72±.02	1.00±.03	0.95±.01	35.5±1.0	98.4±1.6	48.0±0.5
	9.72	6.7	2.0	2.4	6.5	0.72±.02	0.99±.02	0.98±.02	35.5±1.0	96.4±0.9	48.8±0.8
Full-fat Soy flour	2.23	6.9	1.5	1.8	6.4	0.47±.01	0.90±.05	0.93±.03	23.2±0.4	81.1±4.4	68.8±2.2
	2.23	6.9	2.0	2.7	6.4	0.47±.01	0.86±.03	0.89±.02	23.2±0.4	83.9±0.6	66.7±1.6
Defatted Soy flour	3.46	6.8	1.5	2.1	6.3	0.51±.01	0.81±.02	0.89±.03	28.3±0.7	71.5±1.9	71.2±1.1
	3.46	6.8	2.0	2.8	6.2	0.51±.01	0.76±.01	0.79±.00	28.3±0.7	65.1±1.7	66.6±0.4
Soy Protein Concentrate	4.28	7.1	1.5	1.9	6.3	0.46±.01	0.82±.02	0.82±.02	15.2±0.3	80.8±5.7	52.2±0.5
	4.28	7.1	2.0	2.8	6.3	0.46±.01	0.76±.01	0.76±.01	15.2±0.3	68.3±2.7	48.1±1.0
Soy Protein Isolate	1.33	7.0	1.5	2.2	6.3	0.89±.02	0.97±.02	1.02±.02	54.2±2.6	94.3±2.3	95.7±1.9
	1.33	7.0	2.0	3.1	6.2	0.89±.02	0.98±.02	1.01±.00	54.2±2.6	96.5±0.8	93.5±1.6

[1] mg Ca/flask = mg Ca/g dry substrate.
[2] I - initial slurry; Po - initial peptic digest, P = at end of peptic digest and PPa = at end of complete peptic + pancreatic digestion.
[3] Units = mean ± S.D. for 3 flasks.
[4] Units of exchangeability: Ratio, specific activity of the mixture: specific activity of the 18,000 x g supernatant.
[5] Units of solubility (µg Ca in 18,000 x g supernatant/µg in mixture) x 100.

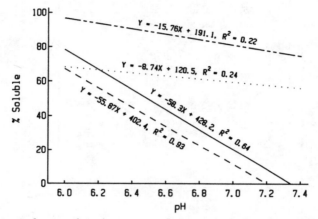

Figure 2. Relationship between calcium solubility and pH after complete digestion for four soy products. Key: solid line, full-fat soy flour; long-dashed line, soy protein isolate; short-dashed line, soy protein concentrate; and dotted line, defatted soy flour.

solubility decreased as pH increased for nearly all foods, the rate of change varied among foods, even in this closely-related series (all soy products).

Several approaches to controlling the pH were tried. These included the use of PIPES, a synthetic buffer with maximum buffering capacity near pH 7 and negligible binding capacity for minerals (12), and the gradual addition of bicarbonate from a dialysis tubing "sack" (5). The method which proved to be most effective was a procedure in which CO_2 was bubbled through the digest mixture both during pH adjustment and throughout the pancreatic digestion (13). With this procedure the pH could be controlled at pH 6.8 ± 0.1.

Enzymes. In an earlier study (14) complete PPa digestion resulted in calcium solubility that was at least 40% higher than when the peptic digestion step was omitted. It was not clear whether the enhanced solubility was due to enzymatic digestion or incubation at acid pH. Therefore, an experiment was run to investigate the relative importance of pH and enzymatic activity. The test substrate was a defatted soy flour. Four minerals, Ca, Mg, Fe and Zn, were measured in the digests. The results are presented in Table II. At the peptic stage, including pepsin increased Fe solubility, but not Ca solubility; Mg

Table II.   Effect of Peptic and Pancreatic Enzymes on
Mineral Solubility After In Vitro Digestion

Digestion	Final pH	% Soluble			
		Ca	Mg	Fe	Zn
pH 2, 60 min. no enzymes	2.1±.0	82.6±1.7	104 ±1.2	2.2±0.7	101.9±5.3
Peptic	2.6±.0	75.3±0.7	98.0±1.1	17.3±0.9	93.4±5.4
pH 2, 60 min. + pH 6.8, 30 min. no enzymes	6.7±.1	41.4±3.6	79.0±3.0	19.3±3.5	56.2±4.6
Peptic + Pancreatic	6.6±.0	48.9±1.2	84.0±1.4	44.7±1.2	56.2±4.6

Note: The substrate was a commercial defatted soy flour product. Values are mean ± S.D. for 4 flasks. Each flask contained 1.0 g substrate. Total mineral content (μg/g substrate) is Ca, 2778; Mg, 3068; Fe, 87; and Zn, 53.

and Zn were completely soluble at pH 2, even without pepsin. A comparison of complete (PPa) digestion with successive non-enzymatic incubations at pH 2 and pH 6.8 indicated that the enzymes greatly increased Fe solubility and slightly increased Ca solubility.

pH and Pancreatic Digestion. The pH range of the in vitro pancreatic digestions (Figure 2) was similar to in vivo conditions, but generally below the pH optima for pancreatic enzymes (10,11). To determine if pH had an indirect effect on calcium solubility through an effect on the rate of digestion, protein and carbohydrate digestion and calcium solubility were measured in the same PPa digests. The digestions consisted of the standard peptic digestion, followed by pancreatic digestion in which the pH was set at 6.2, 6.5, 6.8 or 7.1. The substrates were complex foods, muffins, which had been prepared for another study (15). When 50% of the bran in the bran muffin was replaced by freeze-dried spinach or lettuce, the total calcium in the muffin was increased by 56% or 65%, respectively. The pH had very little effect on carbohydrate or protein digestion, but the calcium solubility dropped by at least 40% in all three muffin formulations over the 0.9 pH unit increase (Table III). Calcium solubility was much lower in the spinach-containing muffins and dropped more sharply with increasing pH than in the bran and lettuce-containing muffins. The calcium data in Table III confirm that a small change in digest pH affects the relative calcium solubilities from different foods.

Bile Salts. We used crude conjugated bile salts mixtures prepared from bile of two species, TC (bovine "sodium taurocholate") and BE (porcine bile extract) to determine if the type of concentration of bile salts affected PPa calcium solubility. Each mixture was used at three concentrations (equal weights), approximately 5, 10 and 15 mM, all within the range found in intestinal contents (11). The substrates were three foods with different lipid compositions: full-fat soy flour, whole milk and whole egg. The "soluble" calcium is aqueous calcium and does not include the non-emulsified lipid layer visible in some of the lower bile salt concentrations. The results in Figure 3 are for digestions at pH 6.8. For soy flour and milk, increasing the concentration of either TC or BE slightly decreased soluble calcium and there were no consistent differences between the two bile salts mixtures. For egg, both mixtures caused a marked decrease in calcium solubility. When the experiment was repeated using a pH maintenance procedure which produced a final pH of pH 6.5, the pattern of results was similar but all solubilities were 5-15% higher.

Fractionation of Calcium from Digests

We have tried two non-destructive approaches to fractionation of the mineral complexes in in vitro digests: ultrafiltration and gel filtration. Only preliminary data from gel filtration experiments are currently available.

An example of the results of ultrafiltration is presented in Table IV. The fractionation of the digest mixture, which was prepared from a soy protein isolate, combines the techniques of centrifugation and ultrafiltration. Since the digest contained bicarbonate as the main buffer, CO_2 was used to apply pressure to the ultrafiltration cell, rather that N_2, to avoid forming precipitates. Most of the digest calcium was soluble after centrifugation at 18,000 x g. Much of it was bound to large complexes, as judged from the filtration of less than half the calcium through the 10,000 MWCO (molecular weight cut-off) membrane. The nominal MWCO of the membrane is not a precise guide to the size of complex which is filtered since one-third of the Ca^{++} from $CaCl_2$ was retained by the 1000 MWCO membrane, but the distribution of calcium in the digest was clearly different from that in the completely ionized solution. For all ultrafiltrations of the in vitro digest, the starting material was the 100,000 x g supernatant in order to avoid confounding the results with sample deterioration. The results were found to be reproducible and not influenced by the

Table III. Effect of In Vitro Pancreatic Digestion pH on Carbohydrate and Protein Digestion and Calcium Solubility

Substrate:[1]	Bran muffin			Spinach/bran muffin			Lettuce/bran muffin		
Soluble component:	maltose[2]	peptide[3]	Ca	maltose	peptide	Ca	maltose	peptide	Ca
	component in 18,000 x g supernatant: mg/g substrate								
Initial. (no digestion)	319	14	(Total= 1.36 mg)	322	24	(Total= 2.12 mg)	328	17	(Total= 2.25 mg)
pH									
6.2	551	108	0.62	542	132	0.28	573	116	0.92
6.5	562	106	0.42	540	130	0.19	556	113	0.83
6.8	551	107	0.37	527	134	0.08	545	116	0.63
7.1	554	108	0.37	517	132	0.06	531	112	0.47

[1] Freeze-dried foods were ground to pass a 40-mesh sieve. Freeze-dried spinach or lettuce replace 50% of the bran of the bran muffins in the spinach/bran and lettuce/bran muffins, respectively.
[2] Maltose, the endproduct of amylase activity, was measured by the method of Dahlquist (16).
[3] Peptide=supernatant protein which was not precipitated when trichloroacetic acid (TCA) solution was added to 5% TCA (W/V) (17).

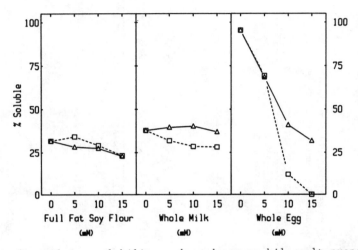

Figure 3.   Calcium solubility: dependence on bile salt preparation
and concentration used in in vitro digestion (pH 6.8–6.9).   Key:
Δ, crude bovine sodium taurocholate; and □, porcine bile extract.

Table IV. Fractionation of Calcium in an In Vitro Digest by
Centrifugation (CE) and Ultrafiltration (UF) and Comparison
with UF Results for $CaCl_2$

Fraction	$CaCl_2$ Solution[3]	In Vitro Digest[4]
	Filterable Calcium, %[5]	
Digest Mixture	–	100%
CE:18,000 x g supernatant[1]	–	87.4±5.3
CE:100,000 x g supernatant[1]	–	67.9±9.1
UF:300,000 MWCO filtrate[2]	–	52.4±6.2
UF:10,000 MWCO filtrate[2]	–	44.9±3.8
UF:5,000 MWCO filtrate[2]	95.8%	–
UF:1,000 MWCO filtrate[2]	67.8	30.2±6.9
Ca recovery (UF)	98%	93–103%

[1]Centrifugation at 18,000 x g, $4°C$ for 15 min. or 100,000 x g, $4°C$
for 60 min.

[2]Ultrafiltration was carried out in an Amicon stirred cell, 65 ml
capacity, at $4°C$ with continuous stirring under CO_2 pressure until
25 ml filtrate had been collected. MWCO = nominal molecular weight
cut-off of the membrane.

[3]50 ml of the solution, containing 50 µmoles Ca, was introduced into
the UF chamber for each determination. Values are for duplicate
experiments.

[4]The in vitro digest was prepared by digesting a soy protein isolate.
For each UF determination, 50 ml of the 100,000 g supernatant was
introduced into the UF chamber. Values are mean ± S.D. for four
experiments.

[5]% filterable calculation: (µg Ca/ml of filtrate)/(µg Ca/ml initial
solution in UF chamber) x 100. The in vitro digest filterable
calcium values were standardized as a % of the original digest mixture calcium.

order of the series of membranes in a particular experiment. Thus, sample deterioration was not a problem when the sample was stored on ice and CO_2 saturation maintained. Recovery of the calcium from the combined filtrate and "retentate" in the ultrafiltration chamber was 93% or better.

Discussion

Exchangeability. Measurement of food mineral absorption in humans by the extrinsic isotope or extrinsic tag method is simpler, more flexible and would allow more efficient use of stable isotopes than other methods for measuring mineral absorption (7), if the relationship between the absorptions of the extrinsic tag and intrinsic food mineral is known. Cook, et al. (18) and Hallberg, et al. (19) originally demonstrated the validity of the extrinsic tag for iron by feeding it together with foods labeled intrinsically with a second isotope. They restricted use of the technique to meals in which there was equal absorption of extrinsic and intrinsic iron, that is, when exchangeability was complete. For other minerals, the requirements for in vivo validation of the extrinsic tag technique, two "safe" isotopes which can be detected in the presence of one another and the laborious process of intrinsically labeling foods, put severe restrictions on the use of the extrinsic tag technique. Also, a lack of information about the degree of exchange between the extrinsic tag and test meal iron has hampered interpretation of results from some iron absorption studies (20).

The in vitro digestion procedure should simplify the use of the extrinsic tag method. Our results indicate that calcium exchangeability can be determined in vitro since the in vivo and in vitro calcium exchangeability values were similar (8). No intrinsic labeling and only one isotope are required. The same procedure provides the information needed to interpret the results of in vivo extrinsic tag studies. Also, it is not necessary to demonstrate complete exchangeability in order to use the extrinsic tag technique for measuring calcium absorption. If exchangeability is known, even if it is not complete it can be used to calculate intrinsic food mineral absorption from the extrinsic tag absorption (8). Hallberg and Bjorn-Rasmussen (6) reached a similar conclusion in their studies on absorption of "contamination" iron. Measurement of calcium exchangeability is further simplified in that only the peptic digestion step of the in vitro procedure is required since exchangeability does not change after that (Table I).

While our data indicate that in vitro digestion provides a simple means to solve some difficult problems of using the extrinsic tag method for measuring calcium absorption, our conclusions are based on a limited number of foods. They would be strengthened if a wider range of foods is tested in direct in vivo - in vitro comparison studies. The test foods should include foods in which exchangeability might not be completed during peptic digestion such as foods with "indigestible" residues that may be altered and release calcium which may be absorbed in the lower intestine.

Bioavailability. In principle, the in vitro procedure provides a relatively fast and inexpensive means to study calcium bioavailability as a characteristic of foods. A knowledge of the chemistry of the foods themselves is not sufficient since changes occur during the process of digestion and the total chemical environment influences the relative binding constants for the various complexes which may be formed (21,22). At the same time, the host characteristics which restrict in vivo absorption and introduce variability not related to food characteristics (1,2,9) are eliminated.

Choice of Potential Bioavailability Criterion. It is usually assumed that calcium must be soluble and probably ionized in order to be available for absorption (9). For the in vitro procedure, as a first approximation we chose calcium solubility after centrifugation at 18,000 x g as the measure of potential bioavailability (Figure 1). We assumed that this would probably overestimate the available calcium and later work based on fractionation might define the bioavailable calcium more precisely. The data in Table IV illustrate how the choice of criterion for "solubility" could affect the in vitro estimate of potential availability, even if in vitro conditions closely resembled in vivo conditions. Since our in vitro criterion unexpectedly underestimated calcium bioavailability for two of the three foods in the direct in vivo - in vitro comparison (8), it was necessary to determine the in vitro digestion conditions which might be limiting solubility before addressing the choice of appropriate criterion.

Digestion Conditions. Peptic conditions were not emphasized since calcium solubility is high at the peptic stage (Table I) and chyme release to the duodenum is more dependent on particle size than completeness of digestion (23,24).

Analyses of the pancreatic digestion conditions indicate that the pH of the pancreatic digest was more important for determining calcium solubility than enzymatic activity (Figure 2, Tables II and III) or bile salts (Figure 3) for most foods tested. In earlier studies (7,8) the pH increased 0.2-0.7 pH units during the pancreatic digestion, presumably due to a combination of bicarbonate decomposition, digestion endproduct release and the variable buffering capacity of the foods. Even though these pH's are in the range found in the small intestine (11), the observed pH drift occured in a range which is critical for calcium solubility (9). The final pH was apparently too high to result in the net effect seen in vivo (8).

"Standard" Digestion Conditions. As a result of the analyses of digestion conditions we have modified our in vitro digestion to start the peptic digestion at pH 2.0 instead of 1.5, and to control the pH at 6.8 in the pancreatic stage by continuous aeration with CO . We also substituted bile extract for taurocholate since, although calcium solubility was similar, other minerals were more soluble in the bile extract-containing digests (13).

The term "standard" is not meant to denote a digestion procedure which should be routinely used to determine potentially available calcium. Since the number of foods tested so far is limited, it will be more useful to think of the "standard" procedure as a set of conditions to be used to see how well we understand food chemistry and calcium solubility in the gastrointestinal environment. It should be used for measuring the relative solubility of calcium from foods and meals, but mostly in the context of comparisons with in vivo results to define factors which require further study.

Lipid-containing Foods. The decreased calcium solubility in the presence of bile salts (Figure 3) suggests factors which require further study. Bile salts have been shown to enhance calcium absorption from sparingly soluble salts (25) but also enhance lipase activity (26). Since no absorption occurs in the in vitro system, the liberated fatty acids could form insoluble calcium soaps (27). In contrast, in vivo calcium absorption is not inhibited by even large amounts of fat in the diet unless there is a pre-existing malabsorption problem (1,2,9). Presumably, lipid hydrolysis products are absorbed from the intestine fast enough to prevent insoluble calcium soap formation. The complex lipids of egg yolk (28), and their possible effects on lipase activity (29) may explain the marked decrease in egg calcium solubility with increasing bile salts concentration. More information is needed to determine if this is a problem in vivo. In any case, these questions should not prevent the use of the in vitro procedure for measuring exchangeability in lipid-containing foods, since exchangeability is determined at the peptic stage, as shown for whole milk and full-fat soy flour in Table I.

Fractionation of Digest Calcium. The in vitro digestion procedure provides a means of producing the starting material for a more detailed study of the calcium complexes in intestinal digests. This may be desirable in a number of situations:
1. To determine how the degree of exchange between an extrinsic isotope and the intrinsic calcium in the food or meal is affected by the method of incorporating the isotope, the calcium source itself or foods fed with it.
2. To determine how controlled manipulation of digestion conditions influences the distribution of calcium among possible ligands from a food or meal.
3. To describe the calcium complexes formed during a standardized digestion for a number of foods and food mixtures, for comparisons, or to test hypotheses relating food components to intestinal calcium complexes.

Fractionation Methods. Ultrafiltration and gel filtration are non-destructive methods which, based on limited experience, can be used for fractionation of mineral complexes from digests. In earlier studies mineral absorption on the gel material was a problem. Lonnerdal (30) introduced a method of treating dextran gels with sodium borohydride in order to eliminate the mineral-binding sites on the gel. In preliminary studies we have recovered more than 90% of Ca, Mg, Fe, Zn and P in samples applied to a borohydride-treated gel column (Sephadex G-50, Pharmacia Fine Chemicals, Piscataway, NJ). Recovery of Ca (Table IV) and Mg, Fe and Zn from ultrafiltration was also good.

In our experience with ultrafiltration, use of CO_2 pressure to force the filterable material through the membrane in the ultrafiltration procedure introduced a possible source of error. The high pressure caused more CO_2 to be dissolved in the digest supernatant and the pH in the chamber decreased to about 6.2. This may have caused a shift of the mineral among ligands (31). It should be possible to formulate a mixture of CO_2 and N_2 to maintain the pH in the chamber, but we have not pursued this.

Gel filtration may be best used to analyze fractions already separated from a digest supernatant by ultrafiltration, as used in a recent study by Sandstrom, et al. (32). A more precise separation of complexes can be obtained with gel filtration, but the size of sample which can be applied is limited. Thus, in many situations, the sample must be concentrated before being applied to the gel column. Either pre-purification or sample concentration could introduce possible shifts in mineral binding which should be understood for proper interpretation of the results (33).

Summary

Use of an in vitro simulated gastrointestinal digestion procedure in calcium bioavailability research has been discussed. Two distinct types of uses were described: (1) measurement of exchangeability to facilitate dietary calcium absorption studies, and (2) study of the fate of food calcium in the gastrointestinal environment with regard to its potential availability for absorption. Ideas have been incorporated from many sources and only limited testing has been possible so far. We have tried to indicate the advantages of such a procedure as well as where more testing of the validity of the ideas for calcium bioavailability research is required.

Acknowledgments

The research was supported in part by grants from NIH Grant 18569 and USDA Cooperative Agreement 58-320A4-9-91.

Literature Cited

1. Allen, L.H. Am. J. Clin. Nutr. 1982, 35, 783-808.
2. Irwin, M.I.; Kienholz, E.W. J. Nutr. 1973, 103, 1019-95.
3. Jacobs, A.; Greenman, D.A. Brit. Med. J. 1969, 1, 673-6.
4. Narasinga Rao, B.S.; Prabhavathi, T. Am. J. Clin. Nutr. 1978, 31, 169-75.
5. Miller, D.D.; Schricker, B.R.; Rasmussen, R.R.; Van Campen, D. Am. J. Clin. Nutr. 1981, 34, 2248-56.
6. Hallberg, L.; Bjorn-Rasmussen, E. Am. J. Clin. Nutr. 1981, 34, 2808-15.
7. Schwartz, R.; Belko, A.Z.; Wien, E.M. J. Nutr. 1982, 112, 497-504.
8. Wien, E.M.; Schwartz, R. J. Nutr. 1983, 113, 388-93.
9. Wilkinson, R. In "Calcium, Phosphate and Magnesium Metabolism"; Nordin, B.E.C., Ed.; Churchill-Livingston: Edinburgh, 1976; pp. 36-112.
10. Harper, H.A. "Review of Physiological Chemistry"; Lange: Los Altos, 1975, 15th ed.; p. 230.
11. Fordtran, J.S.; Locklear, T.W. Am. J. Digest. Dis. 1966, 11, 503-21.
12. Good, N.E.; Winget, G.D.; Winter, W.; Connolly, T.N.; Izawa, S.; Singh, R.M.M. Biochemistry 1966, 5, 467-77.

13. Schwartz, R. 1984, In preparation.
14. Belko, A.Z. M.S. Thesis, Cornell University, Ithaca, 1980.
15. Schwartz, R.; Spencer, H.; Welsh, J.E. Am. J. Clin. Nutr. 1984, 39, 571-6.
16. Dahlquist. A. Scand. J. Clin. Lab. Invest. 1962, 14, 145-51.
17. Layne, E. In "Methods in Enzymology"; Colowick, S. P.; Kaplan, N.O. Eds.; Academic: New York, 1956, Vol. III., pp. 448-50.
18. Cook, J.D.; Layrisse, M.; Martinez-Torres, C.; Walker, R.; Monsen, E.; Finch, C.A. J. Clin. Invest. 1972, 51, 805-15.
19. Hallberg, L.; Bjorn-Rasmussen, E. Scand. J. Haematol. 1972, 9, 193-7.
20. Consaul, H.R.; Lee, K. J. Agr. Food Chem. 1983, 31, 684-9.
21. Leigh, M.J.; Miller, D.D. Am. J. Clin. Nutr. 1983, 38, 202-13.
22. Schubert, J. In "Iron Metabolism"; Gross, F., Ed.; Springer-Verlag: Berlin, 1964; pp. 466-94.
23. Davenport, H.W. "Physiology of the Digestive Tract"; Yearbook Medical Publ.: Chicago, 1982, 5th ed.
24. Arnold, J.G.; Dubois, A. Digest. Dis. Sci. 1983, 28, 737-41.
25. Webling, D. D'A.; Holdworth, E.S. Biochem. J. 1966, 100, 652-60.
26. Rathelot, J.; Julien, R.; Canioni, P.; Coeroli, C.; Sarda, L. Biochemie 1975, 57, 1117-22.
27. Patton, J.S.; Carey, M.C. Science 1979, 204, 145-8.
28. Parkinson, T.L. J. Sci. Fd. Agric. 1966, 17, 101-11.
29. Patton, J.S.; Carey, M.C. Am. J. Physiol. 1981, 241, G328-36.
30. Lonnerdal, B. In "Trace Element Analytical Chemistry in Medicine and Biology"; Bratter, P.; Schramel, P. Eds.; Walter De Gruyter: Berlin, 1980; pp. 439-46.
31. Danielson, B.G.; Pallin, E.; Sohtell, M. Uppsala J. Med. Sci. 1982, 87, 43-53.
32. Sandstrom, B.; Keen, C.L.; Lonnderdal, B. Am. J. Clin. Nutr. 1983, 38, 420-8.
33. Chaberek, S.; Martell, A.E. "Organic Sequestering Agents"; John Wiley: New York, 1959; p. 101.

RECEIVED October 15, 1984

Assaying Calcium Bioavailability in Foods
Applicability of the Rat as a Model

ARTHUR W. MAHONEY and DELOY G. HENDRICKS

Department of Nutrition and Food Sciences, Utah State University, Logan, UT 84322

All middle aged adults lose bone which becomes debilitating when sufficient mineral is lost and fractures occur whether as chronic compression fractures of the vertebrae or as acute fractures of the femoral neck. Evidence is accumulating that adult bone loss is the result of insufficient consumption of bioavailable calcium. Several strategies for assaying calcium bioavailability are discussed. Information is presented supporting the rat as a model for predicting human calcium utilization. This cannot be fully evaluated, however, because animal data have been obtained using growing rats fed controlled amounts of calcium and human data have been obtained from adult subjects who have received liberal amounts of calcium. Calcium absorption data are needed from animal and human subjects having similar nutritional and physiological characteristics and which have consumed identical calcium sources.

Adult bone loss is one of the most debilitating health problems in modern western society for elderly people. Although bone is lost by both men and women as they age (73,95), women suffer from osteoporosis more frequently and severely than do men. Bone loss is detected by radiodensity and photon absorption techniques. Because 20 to 50 percent of bone mineral may be lost before the loss is detected by radiodensity techniques (1,2), it is probable that bone mineral is being lost much earlier than age 40 to 45 in women and age 60 in men as is commonly thought (3,4). Photon absorptiometry has a precision of 2 to 4 percent relative to bone mineral content of the same bone. Measurements on the radii and ulnae are highly correlated (r = 0.85) with bone mineral content of the femoral neck (2). Using photon absorptiometry, Mazess et al. (5) reported that bone mineral declines beginning approximately at age 50 for both men and women. It is estimated that the average rate of this bone loss amounts to approximately 10 mg calcium daily for men and 20 mg calcium for women before menopause. After menopause this loss is

approximately 40 to 120 mg calcium daily (6). Calculating from the
data reported by Mazess et al. (5), approximately 50 mg of bone
mineral is lost daily by women over age 50. It is generally
believed that the larger the bone mass before age-onset bone loss
occurs the less likely the development of debilitating bone loss
after age 65 (3,6,7).

Bone strength however declines much earlier in life beginning
approximately age 20 for both men and women (8). In animals bone
strength is directly related with its mineral content (9-12).
These are related to the amounts of dietary calcium and phosphorus
(11,12) or other factors affecting mineral metabolism (9,10). In
people, however, the ash, calcium and phosphorus concentrations of
bones do not seem to change with age (13) even though human bone
strength (8) and bone mass (3,14) clearly do decline with aging.
Havvi et al. (13) also found descrepancies between bone densities
and radiologically detected osteoporosis vs. bone mineral contents.
In spite of these descrepancies, bone loss is a major difficulty
facing aging people.

Evidence is accumulating for dietary calcium deficiency being
an entity in human nutrition (6,7,15,16). In a study of 130 normal
perimenopausal women, Heany et al. (15) found that their calcium
balance averaged -25 to -30 milligrams daily. By regression
analysis they determined that these women required an intake of
1241 mg (with a 95% confidence interval of 1166 to 1316 mg Ca)
calcium daily to maintain calcium balance. From other data, it is
estimated that 35 mmol (1400 mg) calcium daily is needed to
maintain calcium balance in women aged 35 to 50 and postmenopausal
women need 47.5 mmol (1900 mg) calcium daily (6). All of these
values are well above the current Recommended Dietary Allowance of
800 mg calcium daily for adult women. The average daily calcium
intakes of American women above age 23 vary from 515 to 604
milligrams for different age groups (17). An average daily calcium
intake of 944 (Sd = 343) milligrams was found for 100 premenopausal
Canadian women (18). Calcium is the only nutrient that is
associated with incidence of bone fracture (6). Calcium intake is
highly correlated with the mineral content of the bones of
experimental animals (11). Thus, calcium could be considered the
most frequently deficient nutrient in the U.S.A. Anything that
could result in greater intakes of calcium and/or improved calcium
bioavailability would be potentially important in preventing or
delaying debilitating bone loss in the elderly.

Approximately, 46 percent of all calcium consumed by Americans
is from dairy products (17). Scythes et al. (18) found that dairy
products contributed 66.7 percent of the calcium consumed by
Canadian pre-menopausal women. Others suggest that dairy products
contribute approximately 75 percent of the calcium consumed (19).
Neither data set includes calcium taken as supplements. About nine
percent of the population consume calcium supplements (20).
Dairy products, however, contribute only 13.8 to 18.2 percent of the
energy consumed (17,18). Clearly, dairy products are a rich source
of dietary calcium (approximately 1370 mg per 1000 kcal) and can
contribute major quantities of calcium to the diets of those who
consume them.

Calcium retention is dependent on two factors, absorption and excretion. Normal subjects have been observed to have apparent calcium absorptions of 23 (sd = 12) to 27 (sd = 17) percent of the calcium from normal diets (21,22). For 20 women aged 55 to 65 consuming 629 (se = 92) milligrams dietary calcium daily, the apparent absorption was 32.1 (se = 1.9) percent (23). An apparent calcium absorption of 29.5 percent (n = 130) may be calculated from data published by Heaney et al. (15). Apparent absorption values from 29 to 42 percent may be calculated from data published by Linkswiler (24,25). However, much lower apparent absorption values of 6 to 15 percent may also be calculated from data published from the same laboratory (26). Although there is considerable variability in the apparent absorption values determined from many studies, a conservative value of 25 percent seems realistic for normal people consuming typical diets.

Calcium retention is also affected by variations in urinary excretion. Dietary factors affecting calcium bioavailability have been recently reviewed (19). Linkswiler and her students have shown that dietary protein is a major factor contributing to urinary calcium excretion (24,25,27,28). Renal acid excretion increases with protein intake. Lutz (29) has found that sodium bicarbonate ingestion will alkalinize the urine and reverse the renal excretion of calcium by people treated with a high protein diet. Renal acid secretion and calcuria occur during short-term starvation (30). Ingestion of 5 grams of calcium lactate (650 mg Ca) corrects the acidosis of short-term starvation and improves the calcium balance; however, sodium bicarbonate alone markedly reduces the starvation acidosis but does not improve the calcium balance (30) as it did above (29) for people treated with high protein diet. Thus correction of acidosis does not seem to be the primary factor in controlling urinary calcium excretion.

Dietary phosphorus also affects calcium metabolism. Polyphosphate decreases calcium absorption in young men while orthophosphate supplement does not (26). However, in the rat all forms of phosphate decrease calcium absorption about equally (31). However, widely divergent dietary calcium:phosphorus ratios do not seem to affect calcium utilization by people as long as there is adequate phosphorus intake (32). In general phosphorus stimulates calcium retention in man (32).

Many other dietary factors have been reported to affect calcium bioavailability. Phytate, fiber, cellulose, uronic acids, sodium alginate, oxalate, fat (only in the presence of steatorrhea), and alcohol have been reported to decrease calcium bioavailability (15). Lactose and medium chain triglyceride increase it (15). Fluoride also affects calcium retention primarily by stimulating bone formation thereby decreasing calcium excretion (33-38). The effects of fluoride on calcium utilization have been variable (34,38,39).

Strategies for determining calcium bioavailability

The term bioavailability implies that fraction of a nutrient, drug or toxicant that is utilized relative to the amount consumed. Calcium is fed to the test subject in amounts below what the subject will utilize. This ensures that all of the calcium provided can be absorbed and metabolized. Then, that fraction

which is utilized relative to that given is considered to be the
amount of calcium in the source that is metabolizable.

There are two primary approaches to determining bioavailability: (a) Direct measurement of uptake into the body can be
done using pharmacokinetic methods, quantitating the accumulation of
radioactive nutrients, or by quantitating the accumulation of
a unique mineral or compound above an expected background level. The
uptake into the body can be estimated indirectly by traditional
metabolic balance methods (31,40). One can also use changes in
blood concentrations of minerals, compounds or physiological markers
in conjunction with body weight data to calculate an estimate of
mineral or compound uptake (41-45). (b) The second approach is to
determine the uptake of a test mineral or compound relative to the
uptake of a stable reference source of that mineral or compound
(11,46). Calcium carbonate has frequently been used as a reference
source in animal studies of calcium bioavailability.

Nearly all of the calcium in the body is located in bone.
Bone is very sensitive to dietary factors such as the amount of
calcium present in the diet and the availability of that calcium
when all other nutrients are present in adequate amounts (46, 47).
This is especially true of the growing animal which is utilized in
most bioavailability studies. Adult animals, however, may also be
used. Krook et al (48) caused osteoporosis in adult dogs in 42
weeks by feeding a low-calcium high-phosphorus diet. The bones were
radiologically normal after 28 weeks of calcium repletion (48). The
ash contents of the vertebral bones of these dogs were much more
responsive to dietary calcium and phosphorus manipulation than were
the humeri and femora (48).

The rat appears to be a good animal model that might be
developed for predicting calcium bioavailability for human beings.
Various dietary and physiological factors affect human and rat
calcium absorption similarly (Table I). The greatest discrepancy
among studies seems to be human and rat responses to changes in
dietary phosphorus; increases in dietary phosphorus consistently
decrease calcium absorption by the rat but does not consistently
decrease it in man. However, the calcium absorption response was
similar for rats and humans for 8 of 9 dietary and physiological
factors reviewed (Table I). This is good evidence that the rat may
be a practical model for estimating human dietary calcium
utilization.

An attempt was made to collate data on human and rat apparent
calcium absorption values for several calcium sources. Absorption
values were so variable within species and calcium sources that a
correlation could not be justified. Much of this variability may be
due to methodological differences between the design of the rat and
the human experiments. Most of the animal experiments were
conducted using rapidly growing rats which were fed modest
amounts of calcium but which have high calcium requirements.
On the other hand, most of the human experiments were conducted
using adult subjects consuming liberal amounts of calcium.
Some degree of standardization of methodologies for rats and
human experimentation must be done before a reasonable
comparison can be made on the correlation between the calcium
absorption responses of these two species.

Table I.   Comparison of Various Dietary and Physiological Factors on Apparent Calcium Absorption by Rats and Humans

Factor	Human Beings	Rat	Reference
Ca absorption decreases with age	True		59,60
	True (over age 60)		61
	True (over age 30)		16
		True	62,63
		False	64
Gastric acidity necessary for absorption of poorly soluble Ca source	True		65-67
	True in Billroth II patient		68
			69
	False (Total gastrectomy)	True	9,10,31
Ortho phosphate	No change		26,70-72
	Slight decrease		28
	Decrease		27,73
		Decrease	31,74,75
Polyphosphate	Decrease		26
		Decrease	31
Increased dietary protein	Increase		25,27,76 & 77
		Increased absorptive cap	78
		No change in absorptive cap	79
		No change	81
		Increase	80,82
Intestine adapts to Low CA intake by increasing absorption	True		16,83
		True	84,85,99
Lactose	Increase[a]		68[a],86
	No change		87
		Increase	88
		No change	89-91
Pregnancy	Increase		92-93
		Increase	94,95
Lactation	Increase?		
		Increase	94,96,97

[a] Ca^{-47} uptake in serum.

Although present data cannot be used to determine the
correlation between human and rat calcium absorption responses,
Bressani (49) found a high correlation between nitrogen balance
index in children and nitrogen growth index in rats. Using
published data, a high correlation (r = 0.94) for percent iron
utilization between rats and human beings was found (Mahoney &
Hendricks, unpublished data). Also, iron absorption by rats and
humans responded similarly to 18 of 19 dietary and physiological
factors reviewed which are known to affect iron utilization (Mahoney
& Hendricks, unpublished data). Again, phosphorus seemed to be the
differing factor for these two species. We believe that this
evidence along with that presented in Table I is indicative of the
potential utility of the rat for quantitative preduction human
utilization of many nutrients including calcium. This will require
much concerted research to determine.

  Metabolic Balance Methods. Theoretically, the amount of
mineral retained in the body should be determinable by balance
methods. Heroux and Peter (50) attempted to do this for calcium and
magnesium in rats fed three diets. For rats fed their stock diet,
they predicted from balance data that the carcasses would contain
23.8 g calcium and 605 mg magnesium. By analysis, the carcasses
contained 4.45 g calcium and 152 mg magnesium. However, the
relationship between calcium balance data (X) and carcass data (Y)
were closely related (Y = 1.05X - .03, r = 0.99 for group mean data)
in the rat data of Whittemore et al. (51).

  The metabolic balance technique has received much criticism
(40,50,52-54). The intake and collection errors are usually not
random, the intake usually being slightly overestimated and the
output being slightly underestimated, seldom the reverse (53). As a
result a slight positive error in the balance of nutrients is
usually encountered (50). Thus, alternate methods for determining
the bioavailability of nutrients are sought. The most meaningful
methods, however, will be those that present the bioavailability of
test sources in terms of the amount of the nutrient utilized
relative to that consumed. Certainly, meticulously executed
balance studies will continue to be very valuable for evaluating
nutrient utilization.

  Isotope Methods. The isotopes of calcium have relatively
short half-lives and are readily counted using liquid scintillation
or gamma counters as appropriate to the nuclide. Calcium isotopes
may be quantitated in the excreta, blood, tissues or in the whole
body. This has made them useful for many nutritional metabolic
studies. However, because of safety concerns, radioactive isotopes
are cumbersome to work with and many researchers are unwilling to
administer them to human beings. This has limited the use of
isotopes to those studies in which alternate methods are not
available or are imprecise. Methodologies for stable isotopes of
calcium, which may be safely used in human being, are becoming
available for use in metabolism studies. These will be
practical alternatives to radioactive isotopes in the future.

  Isotopic methods for estimating calcium absorption have been
evaluated by several researchers (49,55-58). From the human data of
Harrison et al. (55), the relationship between percent calcium
absorption determined by isotope dilution (Y) and excreta counting

(X) may be expressed as: Y = 0.83 X + 2.4, r = 0.77. Also from
their data, the relationship between percent calcium absorption by
isotope dilution (Y) and absorption determined by whole body
counting plus urine was: Y = 0.90 X + 3.9, r = 0.96. From the
human data of Agnew et al. (56), the relationship between calcium
retention determined by whole body counting (Y) or by excreta
counting (X) was: Y = 0.87 X - 2.0, r = 0.93. From human data
(58), the relationship between percent calcium absorption determined
by isotope dilution (Y) and absorption determined by
fecal counting (X) was excellent: Y = 0.99 X + 1.04, r = 0.99.
Similarly, the relationship between percent calcium absorption
determined by isotope dilution (Y) and calcium balance (X) was
excellent: Y = 0.92 X - .02, r = 0.97 calculated from rat data
(51). It is clear from the above that calcium retention determined
by balance data consistently are greater than when determined by
carcass analysis or whole body counting; however, there is a good
correlation between the two methods. This is especially true when
the calcium is quantitated by isotope counting in the excreta in
the balance method.

    Relative Calcium Bioavailability. Determination of the
bioavailability of calcium in low-calcium food sources is difficult
using metabolic balance techniques. It can be done, however, using
changes in bone composition in growing animals fed test sources
compared with animals fed a reference source such as calcium
carbonate. A dose-response curve for the reference calcium source
is created by feeding diets containing different amounts of calcium;
and, the bone response is determined. The bone response of the
animals fed the test calcium source may then be compared with the
response expected for an equal dose of calcium from the reference
source. This is done by determining the equation for the linear
portion of the dose-response curve for the reference substance by
the method of least squares. Using this equation, one can then
calculate the response anticipated for the reference substance at
the actual dose of the test substance. Relative bioavailability
(RB) may then be calculated as follows:

$$RB = \frac{\text{Actual Response of test diet}}{\text{Response estimated from CaCO}_3} \times 100$$

    If this model is selected, one must then decide what variables
to use for the ordinate and the abscissa. The parameters must be
dose-sensitive, free of confounding variables, easily determined
and preferably linear. We have evaluated this approach for
estimating the bioavailability of calcium in mechanically deboned
meat products (11). Typically, correlations between various bone
parameters and dietary calcium are very high (r = 0.943 to 0.999).
This is consistent with what others have found for similar
parameters (46,47). These correlations are also similar to the
those (r = 0.947 to 0.982) between the amount of calcium consumed
and calcium retained (11) a good index procedure.

    A very important advantage of the relative bioavailability
assay is that experimental parameters may be selected which are
easily quantitated. Thus, Tso et al. (11) determined the
relationship between dietary calcium concentration (X, g/kg) and

bone weight (Y, mg):   Y = 8.88 X + 66.7, r = 0.99. Relative
bioavailabilities determined with this relationship were similar to
those determined by calcium retained compared with the calcium
consumed or by percent bone ash relative to dietary calcium
concentration. They found that bone breaking strength was also a
good parameter for determining relative bioavailability. Similar
results were found on recalculating the rat data reported by Wong
and LaCroix (57). Using their data, relative calcium
bioavailabilities determined by using bone weight (Y, mg) vs.
dietary calcium (X, g/kg) were similar in two experiments to those
computed by slope-ratio. There was a good correlation (r = 0.92)
between relative bioavailabilities determined by bone weight (Y,
mg) or bone calcium (Y, mg) vs. dietary calcium (X, g/kg). Thus,
relative bioavailabilities are readily determined using easily
quantitated parameters of calcium metabolism. Furthermore,
relative bioavailabilities determined as described above may be
used to evaluate calcium bioavailability in sources having low
calcium concentrations.
     From the foregoing, it is clear that relative
bioavailabilities for various sources may be determined using
easily analyzed parameters of calcium metabolism. Determining the
amount of calcium consumed (X, mg) and the dry weight of some
easily dissected bone (Y, mg) seem logical parameters for
evaluating relative calcium bioavilability. By doing this on an
individual animal basis, one accounts for variations in food
intake and weight gain which is not done when calcium dose is
expressed as concentration in diet or bone. This is particularly
important when evaluating calcium bioavailabilities of food sources
which can cause marked variations in animal acceptance of test
diets. The data may then be analyzed with appropriate statistical
models for determining relative bioavailabilities.
     Relative bioavailability determinations are limited by what is
known about the quantitative absorption of the reference substance.
One cannot confidently predict absorption of the test substances
based on relative bioavailability data. Relative bioavailability
data, however, can be used to rank the test sources and to provide
a basis for comparison among experiments.

Acknowledgments
     Paper 2975 of the Utah State University Agricultural
Experiment Station.

Literature Cited
1. Lutwak, L. J. Am. Dietet. Assoc. 1964, 44, 173-175.
2. Mazess, R.B. In: Barzel, U.S., Ed., "Osteoporosis II", Grune
    & Stratton, Inc., New York, NY, 1979.
3. Garn, S.M. "The Earlier Gain and the Later Loss of Cortical
    Bone in Nutritional Perspective", Charles C. Thomas,
    Publisher, Springfield, IL 1970.
4. Parfitt, A.M. Medical Times 1981, November.
5. Mazess, R.B.; Peppler, W.W.; Chesney, R.W.; Lange, T.A.;
    Lindgren, U.; Smith, E., Jr. Calcif Tissue Int. 1984, 36,
    8-13.
6. Parfitt, A.M. Lancet 1983, 2, 1181-1184.
7. Avioli, L.V. Fed. Proc. 1981, 40, 2418-2422.
8. Lindhal, O. Acta Orthop. Scand. 1976, 47, 11-19.

9. Mahoney, A.W.; Hendricks, D.G. Nutr. Metabol. 1974, 16, 375-382.
10. Mahoney, A.W.; Holbrook, R.S.; Hendricks, D.G. Nutr. Metabol. 1975, 18, 310-317.
11. Tso, N.; McLaughlin, K.; Mahoney, A.W.; Hendricks, D.G. J. Nutr. 1984, 114, 946-952.
12. Crenshaw, T.D.; Peo, E.R., Jr.; Moser, B.D. J. Animal Sci. 1981, 53, 827-835.
13. Havvi, E.; Reshef, A.; Schwartz, A.; Guggenehim, K.; Bernstein, D.S.; Hegsted, D.M.; Stare, F.J. Israel J. Med. Sci. 1971, 7, 1055-1062.
14. Newton-John, H.F.; Morgan, D.B. Clin. Orthop. Rel. Res. 1970, 71, 229-252.
15. Heaney, R.P.; Recker, R.R.; Saville, P.D. Am. J. Clin. Nutr. 1977, 30, 1603-1611.
16. Heaney, R.P.; Gallagher, J.C.; Johnston, C.C.; Neer, R.; Parfitt, A.M.; Chir, M.B.; Whedon, G.D. Am. J. Clin. Nutr. 1982, 36, 986-1013.
17. Nationwide Food Consumption Survey 1977-78. Preliminary Report No. 2. "Food and Nutrient Intakes of Individuals in 1 day in the United States, Spring 1977", SEA/USDA, Washington, D.C., 1980.
18. Scythes, C.A.; Gibson, R.S.; Draper, H.H. Nutr. Res. 1982, 2, 385-396.
19. Allen, L.H. Am. J. Clin. Nutr. 1982, 35, 783-808.
20. Schutz, H.G.; Read, M.; Bendel, R.; Bhalla, V.; Harrill, I.; Sheehan, E.T.; Standal, B.R. Am. J. Clin. Nutr. 1982, 36, 897-901.
21. Kocian, J.; Skala, I.; Bakos, K. Digestion 1973, 9, 317-324.
22. Wilz, D.R.; Gray, R.W.; Dominguez, J.H.; Lemann, J., Jr. Am. J. Clin. Nutr. 1979, 32, 2052-2060.
23. Recker, R.R.; Saville, P.D.; Heaney, R.P. Ann. Int. Med. 1977, 6, 649-655.
24. Anand, C.R.; Linkswiler, H.M. J. Nutr. 1974, 104, 695-700.
25. Walker, R.M.; Linkswiler, H.M. J. Nutr. 1972, 102, 1297-1302.
26. Zemel, M.B.; Linkswiler, H.M. J. Nutr. 1981, 111, 315-324.
27. Hegsted, M.; Schuette, S.A.; Zemel, M.B.; Linkswiler, H.M. J. Nutr. 1981, 111, 553-562.
28. Schuette, S.A.; Linkswiler, H.M. J. Nutr. 1982, 112, 338-349.
29. Lutz, J. Am. J. Clin. Nutr. 1984, 39, 281-288.
30. Kocian, J.; Brodan, V. Nutr. Metab. 1979, 23, 391-398.
31. Mahoney, A.W.; Hendricks, D.G. J. Food Sci. 1978, 43, 1473-1476.
32. Spencer, H.; Kramer, L.; Osis, D. Am. J. Clin. 1982, 36, 776-787.
33. Jowsey, J.; Riggs, B.L.; Kelly, P.J.; Hoffman, D.L. Am. J. Med. 1972, 53, 43-49.
34. Cohn, S.H.; Donbrowski, C.S.; Hauser, W.; Atkins, H.L. Am. J. Clin. Nutr. 1971, 24, 20-28.
35. Spencer, H.; Lewin, I.; Osis, D.; Samachson, J. Am. J. Med. 1970, 49, 814-822.
36. Chang, Y.O.; Pan, M.; Varnell, T. Nutr. Rpts. Intl. 1977, 16, 539-547.
37. Ericsson, Y. Calc. Tiss. Res. 1972, 9, 39-53.

38. Rich, C.; Ensinck, J.; Ivanovich, P. J. Clin. Invest. 1964,
     43, 545-556.
39. Riggs, B.L.; Hodgson, S.F.; Hoffman, D.L.; Kelly, P.J.;
     Johnson, K.A.; Taves, D. JAMA 1980, 243, 446-449.
40. Lentner, C.; Lauffenburger, T.; Guncago, J.; Dambachner, M.;
     Haas, H.G. Metabolism 1975, 24, 461-471.
41. Mahoney, A.W.; Van Orden, C.C.; Hendricks, D.G. Nutr. Metab.
     1974, 17, 223-230.
42. Reizenstein, P. Br. J. Haematol. 1975, 31, 265-268.
43. Saarinen, U.M.; Siimes, M.A. Pediatric Res. 1979, 13, 143-147.
44. Adolph, W.H.; Wang, C.H.; Smith, A.H. J. Nutr. 1938, 16, 291-
     297.
45. Pennell, M.D.; Davies, M.I.; Rasper, J.; Motzok, I. J. Nutr.
     1976, 106, 265-274.
46. Ranhotra, G.S.; Gelroth, J.A.; Torrence, F.A.; Bock, M.A.;
     Winterringer, G.L. J. Nutr. 1981, 111, 2081-2086.
47. Forbes, R.M.; Weingartner, K.E.; Parker, H.M.; Bell, R.R.;
     Erdman, J.W., Jr. J. Nutr. 1979, 109, 1652-1650.
48. Krook, L.; Lutwak, L.; Henrikson, P.; Kallfelz, F.; Hirsch,
     C.; Romanus, B.; BeLanger, F.; Marier, J.R.; Sheffy, B.E. J.
     Nutr. 1971, 101, 233-246.
49. Bressani, R. In: Bodwell, C.E., ed., "Evaluation of Proteins
     for Humans", AVI, Inc., Westport, CN 1977, pp. 81-118.
50. Heroux, O.; Peter, D. J. Nutr. 1975, 105, 1157-1167.
51. Whittemore, C.T.; Thompson, A.; Atherton, D. Br. J. Nutr.
     1973, 30, 425-436.
52. Hegsted, D.M. J. Nutr. 1976, 106, 307-311.
53. Forbes, G.B. Nutr. Rev. 1973, 31, 297-300.
54. Ferretti, J.L.; Bazan, J.L.; Puche, R.C. Medicina (Buenos
     Aires) 1976, 36, 83-92.
55. Harrison, J.E.; McNeill, K.G.; Wilson, D.R.; Oreopoulos, D.G.;
     Krondl, A.; Finlay, J.M. Clin. Biochem. 1973, 6, 237-235.
56. Agnew, J.E.; Kehayoglou, A.K.; Holsworth, C.D. Gut 1969, 10,
     590-597.
57. Wong, N.P.; LaCroix, D.E. Nutr. Rpts. Intl. 1980, 21, 673-680.
58. DeGrazia, J.A.; Ivanovich, P.; Fellows, H.; Rich, C. J. Lab.
     Clin. Med. 1965, 66, 822-829.
59. Ireland, P.; Fordtran, J.S. J. Clin. Invest. 1973, 52, 2672-
     2681.
60. Alevizaki, C.C.; Ikkos, D.G.; Singhelakis, P. J. Nuclear Med.
     1973, 14, 760-762.
61. Bullamore, J.R.; Gallagher, J.C.; Wilkinson, R.; Nordin,
     B.E.C.; Marshall, D.H. Lancet 1970, 2, 535-537.
62. Armbrecht, H.J.; Zenser, T.V.; Bruns, M.E.H.; Davis, B.B.
     Am. J. Physiol. 1979, 236, E769-E774.
63. Armbrecht, H.J.; Gross, C.J.; Zenser, T.V. Arch. Biochem.
     Biophys. 1981, 210, 179-185.
64. Hironaka, R.; Draer, H.H.; Kastelic, J. J. Nutr. 1960, 71,
     356-360.
65. Hunt, J.N.; Johnson, C. Digest. Dis. & Sci. 1983, 28, 417-421.
66. Ivanovich, P.; Fellows, H.; Rich, C. Ann. Intern. Med. 1967,
     66, 917-923.
67. Nicolaysen, R.; Ragard, R. Scand. J. Clin. Lab. Invest. 1955,
     7, 298-299.

68. Kocian, J.; Brodan, V. Digestion 1975, 12, 193-200.
69. Bradley, E.L., III,; Isaacs, J.; Hersh, T.; Dvidson, E.D.; Millikan, W. Ann. Surgery 1975, 182, 415-429.
70. Kim, Y. "Effect of Level of Protein, Calcium and Phosphorus Intake on Calcium, Phosphorus and Magnesium Metabolism in the Young Adult Male", Ph.D. Dissertation, University of Wisconsin, Madison, 1977.
71. Spencer, H.; Kramer, L.; Osis, D.; Norris, C. J. Nutr. 1978, 108, 447-457.
72. Malm, O.J. Scand. J. Clin. Lab. Invest. 1953, 5, 75-84.
73. Leichsenring, J.M.; Norris, L.M.; Lamison, S.A.; Wilson, E.D.; Patton, M.B. J. Nutr. 1951, 45, 407-418.
74. Tanaka, Y.; Frank, H.; DeLuca, H. Science 1973, 181, 564-566.
75. Shah, B.G.; Meranger, J.C. Can. J. Physiol. Pharm. 1970, 48, 675-680.
76. Schwartz, R.; Woodcock, N.A.; Blakely, J.D.; MacKellar, I. Am. J. Clin. Nutr. 1973, 26, 519-523.
77. McCance, R.A.; Widdowson, E.M.; Lehmann, H. Biochem. J. 1942, 36, 686-691.
78. Moyer, G.L.; Wilson, H.D.; Schedl, H.P. Am. J. Digest. Dis. 1978, 23, 545-549.
79. Shenolikar, I.S. Nutr. Metabol. 1974, 16, 10-14.
80. Bottom, J.S. "The Effect of High Protein, Low Pyridoxine Diet on Calcium Retention in Rats", M.S. Thesis, Utah State University, Logan, Utah, 1978.
81. Howe, J.C.; Beecher, G.R. Nutr. Rpt. Intl. 1981, 24, 919-929.
82. Engstrom, G.W.; DeLuca, H.F. J. Nutr. 1963, 81, 218-222.
83. Irving, J.T. "Calcium and Phosphorus Metabolism", Academic Press, NY, 1973, p. 40.
84. Kimberg, D.V.; Schachter, D.; Schenker, H. Am. J. Physiol. 1961, 200, 1256-1262.
85. Petith, M.M.; Schedl, H.P. Am. J. Physiol. 1976, 231, 865-871.
86. Kabayashi, A.; Kawai, S.; Ohbe, Y.; Nagashima, Y. Am. J. Clin. Nutr. 1975, 28, 681-683.
87. Mills, R.; Breiter, H.; Kempster, E.; McKey, B.; Pickens, M.; Outhouse, J. J. Nutr. 1940, 20, 467-476.
88. Forbes, R.M. J. Nutr. 1964, 83, 225-233.
89. Ali, R.; Evans, J.L. J. Nutr. 1967, 93, 273-279.
90. Evans, J.L.; Ali, R. J. Nutr. 1967, 92, 417-424.
91. Urban, E.; Pena, M. Digestion 1977, 15, 18-27.
92. Shenolikar, I.S. Am. J. Clin. Nutr. 1970, 23, 63-67.
93. Pitkin, R.M. Am. J. Obstet. Gynecol. 1975, 121, 724-737.
94. Halloran, B.P.; DeLuca, H.F. Am. J. Physiol. 1980, 239, E64-E68.
95. Graves, K.L.; Wolinsky, I. J. Nutr. 1980, 110, 2420-2432.
96. Boass, A.; Toverud, S.U.; Pike, J.W.; Haussler, M.R. Endocrinology 1981, 109, 900-907.
97. Kostial, K.; Gruden, N.; Durakovic, A. Calc. Tiss. Res. 1969, 4, 13-19.
98. Hansard, S.L.; Crowder, H.M. J. Nutr. 1957, 62, 325-339.
99. Krawitt, E.L.; Stubbert, P.A.; Ennis, P.H. Am. J. Physiol. 1973, 224, 548-551.

RECEIVED January 19, 1985

Phosphates and Calcium Utilization in Humans

MICHAEL B. ZEMEL

Department of Nutrition and Food Science, Wayne State University, Detroit, MI 48202

High phosphate diets cause decreased Ca absorption, secondary hyperparathyroidism, accelerated bone resorption and soft tissue calcification in some animals, but not in normal humans. Although phosphates may decrease Ca absorption in man at very high (> 2000 mg/day) Ca intakes, they do not do so at more moderate Ca levels and enhance Ca absorption at very low levels (< 500 mg/day). Phosphates increase renal tubular reabsorption and net retention of Ca. At low Ca intakes, phosphates stimulate parathyroid hormone (PTH) secretion without causing net bone resorption. Increasing both dietary Ca and P causes a decrease in PTH-mediated bone resorption; polyphosphates and phosphorus in food cause greater reductions than does inorganic orthophosphate, as these sources are slowly released in digestion.

The effects of phosphates on the absorption and utilization of calcium have been the topic of continuing controversy. Phosphates have been widely believed to reduce calcium absorption as a result of the formation of insoluble calcium phosphate salts in the gut. However, data from several human studies do not support this conclusion. Although phosphate supplementation may cause decreases in calcium absorption in man at very high levels of calcium intake, i.e. 2000 mg per day or more, it does not appear to do so when calcium intake is 1500 mg per day or less (1,2). Furthermore, Fox and Care (3) found inorganic phosphate to be without effect on calcium absorption from Thiry-Vella jejunal loops in pigs. In addition, phosphates do not consistently cause precipitation of calcium from foods during in vitro digestions (4). Table I shows the effects of 1% ortho-, tri-poly-, or hexametaphosphate added to ground beef or soy protein concentrate on calcium solubility following simulated gastric and complete gastrointestinal digestions. Gastric digestions were accomplished by incubating samples in a pepsin-HCl mixture for two hours at 37°C, and "complete" gastrointestinal digestions were accomplished by raising the pH of the samples to 7.0 with sodium

0097-6156/85/0275-0029$06.00/0

Table I. Effects of 1% Orthophosphate, Tripolyphosphate, and
Hexametaphosphate on Calcium Solubility from Ground Beef
or Soy Protein Concentrate Subjected to In Vitro Gastric
and Gastrointestinal Digestions

	Soluble Calcium (% of Control)	
	Calcium Source	
Gastric Digestion	Soy Concentration	Ground Beef
Control	100.0^a	100.0^a
+orthophosphate	73.4^b	41.0^b
+tripolyphosphate	73.2^b	43.6^b
+hexametaphosphate	72.9^b	38.5^b
Gastrointestinal Digestion		
Control	100.0^a	100.0^a
+orthophosphate	85.0^b	145.5^b
+tripolyphosphate	97.4^a	427.7^c
+hexametaphosphate	127.8^c	400.0^c

Note: Nonmatching superscripts in each column denote significant
($p < 0.05$) differences for each treatment.

bicarbonate and incubating for two additional hours with pancreatic
and bile extracts.
Although all three phosphates tested caused substantial decreases in
calcium solubility following the gastric digestion, this trend was
reversed upon completion of the second phase of the digestion. Only
orthophosphate caused a decrease in calcium solubility from the soy
following the complete digestion, while all three phosphates enhanced
calcium solubility from the meat. Thus, phosphates are unlikely to
interfere with calcium absorption due to the formation of insoluble
salts in the gut.
    In contrast, recent human studies indicate that increasing
dietary phosphorus, in the form of orthophosphate, may cause an
increase in the apparent absorption of calcium (5,6). Table II shows
the effects of 1 g supplements of phosphorus on calcium absorption
during two experiments with 18 young adult males maintained at low
(< 400 mg/day) levels of calcium intake; in both experiments, the
phosphorus supplement caused a significant reduction in fecal calcium. However, in a third study, which employed a slightly higher level
(500 mg/day) of calcium intake, orthophosphate supplementation was
without effect on calcium absorption (7)   Thus, available evidence
indicates that orthophosphates may enhance calcium absorption in
man at very low levels of calcium intake (< 400 mg/day), reduce it
at very high levels of intake ($\geq$ 2000 mg/day), but exert no effect
on calcium absorption throughout a broad range of moderate calcium
intakes (500-1500 mg/day). Consequently, if phosphates are to
affect calcium utilization, they must do so post-absorptively.

Table II. Effects of Orthophosphate on Calcium Absorption

Calcium Intake	Phosphorus Intake	Fecal Calcium	Apparent Absorption Of Calcium
		mg/day	
		Study 1	
399^a	835^a	401^a	- 2
399^a	1835^b	374^b	+25
		Study 2	
356^a	866^a	329^a	+27
356^a	1866^b	284^b	+72

Note: Nonmatching superscripts in each column denote significant differences ($p < 0.05$) for each study.

Phosphates and Bone Mineralization

High phosphate diets low in calcium have been shown to cause bone demineralization and soft tissue calcification in several lower animal species (8-13), and Draper and co-workers (14-16) found similar effects in rats and mice fed diets containing adequate or high levels of calcium. Increasing the level of phosphorus intake from 0.6 to 1.2% resulted in an accelerated loss of ^{45}Ca from deeply labelled bone and a corresponding decrease in the calcium and phosphorus content of the skeleton; these effects were eliminated by parathyroidectomy. The authors concluded that excess dietary phosphorus stimulates parathyroid hormone (PTH) mediated bone resorption. However, in tissue culture, inorganic phosphate inhibits PTH-induced bone resorption (17) and instead stimulates increased bone collagen synthesis (18) and mineralization (19). In addition, Anderson et al. (20) reported that monkeys fed high phosphate diets low in calcium (0.3% Ca) or adequate in calcium (0.95% Ca) developed no significant clinical, radiographic, or histological evidence of bone disease during a 7-year observation period.

In attempting to reconcile these findings, it should be pointed out that rats may not be appropriate models for the study of calcium metabolism in humans. Unlike humans, the rat does not undergo epiphyseal plate closure and does not have a significant haversian remodeling sequence (21). Furthermore, rats excrete only 1-2% of their calcium intake in their urine whereas humans excrete approximately 20-30% or more. This fact is especially significant, since most of the known effects of phosphates on calcium retention in humans are effected by alterations in urinary calcium.

Increasing the level of phosphorus intake has long been known to exert a hypocalciuretic effect (22,23). Bell et al. (24) suggested that this effect may be secondary to an increase in PTH secretion, but phosphorus supplementation has been reported to decrease urinary calcium even in parathyroidectomized rats (14). Goldsmith and co-workers (25) found phosphate supplements to cause an increase in PTH secretion in women with osteoporosis; however, ileac crest biopsies from the women showed increases in both bone formating and bone sorbing surfaces. It was suggested that the phosphate initially caused retention of calcium by the skeleton and a subsequent reduction in serum calcium levels; this reduction in serum calcium was presumed to be responsible for the increase in PTH secretion. It was further suggested that providing a calcium supplement along with the phosphorus may prevent the phosphate-induced decrease in serum calcium and the resultant increase in PTH-mediated bone resorption; thus net bone mineral accretion would be permitted.

Orthophosphates

We have conducted two human metabolic studies (5,6) to compare the effects of increasing phosphorus intake on calcium utilization in healthy young adults maintained at low (ca. 400 mg/day) and high (ca. 1200 mg/day) levels of calcium intake. Increasing dietary phosphorus, as orthophosphate, caused a slight reduction in fecal calcium and a substantial reduction in urinary calcium losses (Table III).

Table III. Absorption, Excretion, and Retention of Calcium as Affected by Calcium and Orthophosphates

Diet	Intake	Apparent Absorption	Urinary Excretion	Calcium Balance
		mg/day		
Low Ca Low P	399[a]	− 2[a]	196[a]	−198[a]
Low Ca High P(ortho)	399[a]	+ 25[b]	115[b]	− 90[b]
High Ca High P(ortho)	1194[b]	+177[c]	176[c]	+ 1[c]

Note: Nonmatching superscripts in each column denote significant differences (P < 0.05).

Consequently, calcium balance improved, although negative calcium balance was still evident in all subjects. Simultaneously increasing calcium and phosphorus intake, however, caused an increase in the apparent absorption of calcium and a decrease in urinary calcium; these changes were sufficient to attain calcium balance. An indication of the mechanism of the hypocalciuretic effect of phosphorus is presented in Table IV. The phosphate supplement caused an increase in fractional renal tubular reabsorption of calcium without affecting glomerular filtration rate; even the combined supplements of calcium and phosphate caused an increase in calcium reabsorption, albeit a smaller one than that oberseved with the low calcium-high phosphorus diet. Diet was without statistically significant effect on plasma ultrafiltrable calcium levels. However, the phosphate supplement did cause a decrease in mean ultrafiltrable calcium levels

at the low level of calcium intake, while no such decrease was observed when calcium and phosphate supplements were given simultaneously. Furthermore, the phosphate supplement caused an increase in
urinary cyclic AMP excretion, indicating an increase in PTH secretion,
while simultaneously supplementing the diet with calcium and phosphate caused reductions in the urinary excretion of both cyclic
AMP and   hydroxyproline, indicating a decrease in PTH-mediated bone
resorption. Thus, it appears that a high calcium-high phosphorus
diet may provide a mechanism to increase net skeletal calcium retention without stimulating PTH release.

Table IV.   Renal Handling of Calcium as Affected by Calcium
and Orthophosphates

Diet	Fractional Renal Ca Reabsorption (%)	Plasma Ultra-filtrable Ca (mg/100 ml)	Urinary Cyclic AMP (µmoles/day)	Urinary Hydroxyproline (mg/day)
Low Ca Low P	97.9[a]	5.53[a]	2.14[a]	27.72[a]
Low Ca High P(ortho)	89.7[b]	5.13[a]	3.22[b]	24.50[a]
High Ca High P(ortho)	98.1[c]	5.52[a]	1.64[c]	16.76[b]

Note: Nonmatching superscripts in each column denote significant
($p < 0.05$) differences.

It is of interest to note that changes in calcium balance were
paralleled by changes in phosphorus balance (Table V). Despite gen-
erous phosphorus intakes (in excess of the RDA) and approximately
60-70% absorption, subjects were in negative phosphorus balance when
calcium intake was low. Phosphorus balance became positive only when
calcium balance was positive, and there was a significant positive
correlation between the two. These data, along with the aforementioned changes in cyclic AMP and hydroxyproline excretion (Table IV) serve
to reinforce the concept that dietary calcium and phosphorus exert
mutually beneficial effects, as the two elements are retained together in the skeleton. This concept is also supported by survey data.
In a cross-sectional study of bone status and fracture rates in two
regions of Yugoslavia, Matkovic et al. (26) reported that metacarpal
cortical width, cortical area and cortical density were all lower
and the fracture rate was higher in the region with the lower calcium intake than in the region with the higher calcium intake. The
differences in density tended to diminish past age 55 in the women,
but was still evident even at age 75. Although the authors attributed these effects to differences in calcium intake between the two
regions, inspection of their data indicates a two-fold increase in
phosphorus intake as well.

Table V. Absorption, Excretion, and Retention of Phosphorus
as Affected by Calcium and Orthophosphates

Diet	Intake	Apparent Absorption	Urinary Excretion	Retention
		mg/day		
Low Ca Low P	835[a]	486[a]	679[a]	-193[a]
Low Ca High P(ortho)	1835[b]	1307[b]	1416[b]	-109[b]
High Ca High P(ortho)	1835[b]	1239[b]	1197[a]	+ 42[c]

Note: Nonmatching superscripts in each column denote significant
(p < 0.05) differences.

Polyphosphates

Condensed (poly) phosphates may exert different effects on calcium
utilization than the aforementioned effects of simple (ortho-) phosphates. Polyphosphates have a much greater affinity for calcium than
do orthophosphates, and soluble calcium-polyphosphate complexes are
readily formed in the gastric and intestinal environments. In addition, polyphosphates must be hydrolyzed by an intestinal alkaline
phosphatase (27) prior to absorption. We have found polyphosphates
to be incompletely (80.5%) hydrolyzed to orthophosphate during the
digestive process in young adult males when calcium intake was low;
only 56% of a 1 g phosphorus supplement was absorbed from a polyphosphate sources as compared to 71% from an orthophosphate source (5).
     It is possible that, by virtue of their incomplete hydrolysis in
the gastrointestinal tract, polyphosphates may represent a "slow-release" form of phosphorus that does not elicit as great a PTH response as orthophosphate. A 75-day repeated measures human metabolic study was recently conducted to compare the effects of 1 g
phosphorus supplements, given either an orthophosphate or as hexa -
metaphosphate, on calcium and phosphorus utilization in ten young
adult males consuming a low (356 mg/day) or a high (1166 mg/day) level of calcium (6). Polyphosphate hydrolysis (Table VI) was found to be
reduced when the level of calcium intake increased. Hexametaphosphate and all of its detectable hydrolysis products, except for
pyrophosphate,were stabilized against further hydrolysis to orthophosphate. Consequently, the fraction of fecal phosphorus in the
form of polyphosphate increased from 62.5% on the low calcium, high
polyphosphate diet to 74.3% on the high calcium, high polyphosphate
diet. As a result of this increase in residual fecal polyphosphate,
the hexametaphosphate caused a greater increase in fecal calcium
losses when calcium intake was high than when it was low (Table VII).
In contrast to orthophosphate, hexametaphosphate caused a decrease in
calcium absorption, as indicated by an increase in fecal calcium, at
both levels of calcium intake. However, the polyphosphate supplement

Table VI. Distribution of Fecal Phosphates as Affected by Level
of Calcium Intake

Phosphate Species	Diet	
	Low Calcium High Polyphosphate	High Calcium High Polyphosphate
ortho	37.6	25.7*
pyro	26.5	27.6*
tripoly	9.1	13.2*
trimeta	2.5	4.0*
tetra	3.9	6.3*
hexameta	20.5	23.2*
total poly	62.5	74.3*

Note: Expressed as a percentage of total fecal phosphorus.

*indicates a significant difference ($p < 0.01$) in each row.

Table VII. Calcium Excretion and Retention as Affected by
Orthophosphate, Hexametaphosphate (Polyphospate), and Calcium

Diet	Calcium Intake	Fecal Excretion	Urinary Excretion	Balance
Low Ca- Low P	356[a]	329[a]	137[a]	-110[a]
Low Ca- High P(ortho)	356[a]	284[b]	101[b]	- 29[b]
Low Ca- High P(poly)	356[a]	338[c]	116[c]	- 98[a]
High Ca- High P(ortho)	1166[b]	832[d]	158[d]	+176[c]
High Ca- High P(poly)	1166[b]	959[e]	169[e]	+ 38[d]

Note: Nonmatching superscripts in each column denote significant
($p < 0.05$) differences.

caused a 54 mg/day increase in fecal calcium when compared to orthophosphate at the lower level of calcium intake, while at the higher
level of calcium intake a much larger increase of 127 mg/day was
observed.

All subjects were in negative calcium balance when consuming the
basal low calcium, low phosphorus diet (Table VII), the mean calcium
loss being 110 mg/day. The orthophosphate supplement significantly
reduced this loss to 29 mg/day, due to decreases in both urinary and
fecal calcium losses. The polyphosphate supplement, however, caused

an increase in fecal calcium and was not as hypocalciuretic as the
orthophosphate; as a consequence it did not improve calcium balance.
Simultaneously increasing the levels of calcium and orthophosphate
intake resulted in net retention of calcium for all subjects. The
improvement in calcium balance compared to that seen when orthophosphate was given without a calcium supplement was due to a substantial
increase in the apparent absorption of calcium which was accompanied
by only a slight increase in urinary calcium losses. The high calcium, high polyphosphate diet caused significantly greater urinary
and fecal losses of calcium than did the high calcium, high orthophosphate diet; nonetheless, calcium balance was attained on both
high calcium diets.

As with the first study (5), calcium and phosphorus balances
varied together, and there was a significant positive correlation
between the two. Phosphorus balance became positive only when calcium
balance was positive, during the two high calcium treatment periods
(Table VIII). The calcium supplement caused a significant increase in
fecal phosphorus losses, but this was more than compensated for by
the substantial decrease in urinary phosphorus caused by the calcium
supplements. Calcium has been shown to affect the renal handling of
phosphorus through both PTH-dependent and PTH independent mechanisms
(28-30). Calcium causes a direct increase in renal tubular reabsorption of phosphorus (29) and inhibits the PTH-dependent renal adenylate cyclase (30). However, the decreases in urinary cyclic AMP
(Table 9) which resulted from calcium supplementation in the present
study suggest a PTH-dependent mechanism.

It is noteworthy that although phosphorus absorption was lower
on the high calcium, high polyphosphate diet than on the high calcium,
high orthophosphate diet, there was no significant difference between

Table VIII. Phosphorus Excretion and Retention as Affected by
Orthophosphate, Hexametaphosphate (Polyphosphate), and Calcium

Diet	Phosphorus Intake	Fecal Excretion	Urinary Excretion	Balance
	mg/day			
Low Ca– Low P	866[a]	376[a]	711[a]	−221[a]
Low Ca– High P(ortho)	1866[b]	437[a]	1450[b]	− 21[b]
Low Ca– High P(poly)	1866[b]	634[b]	1343[c]	−111[a]
High Ca– High P(ortho)	1866[b]	522[c]	1289[d]	+ 55[c]
High Ca– High P(poly)	1866[b]	799[d]	1037[c]	+ 30[c]

Note: Nonmatching superscripts in each column denote significant
(p< 0.05) differences.

the effects of the two diets on phosphorus retention; this was due to
the substantial decrease in urinary phosphorus on the high calcium,
high polyphosphate diet from that observed on the high calcium-high
orthophosphate diet. Thus, phosphorus absorbed from the calcium-polyphosphate supplement was utilized more efficiently than that absorbed
from the calcium-orthophosphate supplement. Similarly, Schuette and
Linkswiler (31) reported lower utilization of phosphorus when an
orthophosphate supplement was the source than when foods were the
phosphorus source; the lower utilization was due to an increase in
urinary phosphorus excretion. The authors noted that, since phosphorus is absorbed as inorganic phosphate, a phosphorus supplement
(i.e. orthophosphate) will be rapidly absorbed and cause an immediate
rise in circulating phosphorus levels and subsequent spillage into
urine. It therefore appears that those phosphorus sources which are
slowly absorbed, such as polyphosphates, may be utilized more efficiently. This concept is supported by observed changes in urinary
cyclic AMP in response to varying dietary orthophosphate, polyphosphate and calcium (Table IX). Cyclic AMP excretion increased when the
phosphate supplements were given without calcium and the greater increase was associated with the orthophosphate supplement. In contrast, simultaneously increasing the levels of calcium and phosphorus
intakes caused a decrease in urinary cyclic AMP, and the largest decrease resulted from the combined calcium-polyphosphate supplement.
Furthermore, the calcium-polyphosphate supplement also caused a decrease in bone resorption, as indicated by a significant decrease in
hydroxyproline excretion. Thus these data indicate that, in the
presence of adequate calcium, polyphosphates act as a "slow-release"

Table IX. Urinary Cyclic AMP and Hydroxyproline as Affected by
Orthophosphate, Hexametaphosphate, and Calcium

Diet	Urinary Cyclic AMP[1] (μmoles/day)	Urinary Hydroxyproline[2] (mg/day)
Low Ca Low P	4.32^a	27.10^a
Low Ca- High P(ortho)	4.71^b	25.82^a
Low Ca- High P(poly)	4.53^a	28.14^a
High Ca- High P(ortho)	4.22^a	31.03^a
High Ca- High P(poly)	4.10^c	24.80^b

[1] No significant differences were observed at $\alpha = 0.05$; nonmatching
superscripts in this column denote differences at $\alpha = 0.06$.

[2] Nonmatching superscripts in this column denote significant
($p < 0.05$) differences.

source of phosphorus which allows for net deposition of bone mineral without depressing circulating calcium levels and, consequently, without stimulating PTH secretion.

Conclusions

Data presented herein indicate that although orthophosphates cause a slight increase in calcium absorption when calcium intake is low and polyphosphates cause a slight decrease, phosphates exert little significant effect on calcium absorption throughout a broad range (500 - 1500 mg/day) of normal calcium intakes. Phosphates do exert a significant hypocalciuretic effect and thereby cause an improvement in calcium retention. However, at very low levels of calcium intake phosphates may reduce circulating calcium levels and , as a consequence,elicit an increase in parathyroid hormone-mediated bone resorption which partially offsets the improvement in calcium retention. On the other hand, high levels of calcium and phosphorus intakes are associated with substantial retention of both minerals and a decrease in parathyroid hormone-mediated bone resorption. These results also indicate that the effects of dietary phosphates on calcium and phosphorus utilization in man is in part dependent upon the phosphate source. Those phosphates which are slowly digested and absorbed, such as food-bound phosphates or a polyphosphate, appear to be more efficiently utilized then inorganic orthophosphate.

Literature Cited

1. Spencer, H.; Kramer, L; Osis, D.; Norris, C. J. Nutr. 1978, 108, 447-457.
2. Kim, Y.; Linkswiler, H.M. Fed. Proc. 1980, 39, 895.
3. Fox, J.; Care, A.D. Br. J. Nutr. 1978, 39, 431-439.
4. Zemel, M.B. Unpublished Data.
5. Zemel, M.B.; Linkswiler, H.M. J. Nutr. 1981, 111, 315-324.
6. Zemel, M.B.; Soullier, B.A.; Steinhardt, N.J. Fed. Proc. 1983, 42, 397.
7. Hegsted, M.; Schuette, S.A.; Zemel, M.B.; Linkswiler, H.M. J. Nutr. 1981, 111, 553-562,
8. Krook, L.; Barret, R.B.; Usui, K.; Wolke, R.E. Cornell Vet. 1963, 224-240.
9. Rowland, R.N.; Capen, C.C.; Nagode, L.A. Path. Vet. 1968, 5, 504-519.
10. Hammond, R.H.; Storey, E. Calc. Tiss. Res. 1970, 4, 291-304.
11. Morris, M.L. Jr.; Teeter, S.M.; Collins, D.R. J. Am. Vet. Med. Assoc. 1971, 477-488.
12. Laflamme, G.H.; Jowsey, J. J. Clin. Invest. 1972, 51, 2834-2840.
13. Jowsey, J.; Reiss, E.; Canterbury, J.M. Acta Orthop. Scand. 1974, 45, 801-808.
14. Anderson, G.H.; Draper, H.H. J. Nutr. 1972, 102, 1123-1132.
15. Draper, H.H.; Sie, T.L.; Bergen, J.G. J. Nutr. 1972, 102, 1133-1142.
16. Krishnarao, G.V.G.; Draper, H.H. J. Nutr. 1972, 102, 1143-1146.
17. Raisz, L.G.; Neimann, I. Endocrinology 1969, 85, 446-452.
18. Raisz, L.G. Fed. Proc. 1970, 29, 1176-1178.
19. Flanagan, B.; Nichols, G. Jr. J. Clin. Invest. 1969, 48, 607-612.

20. Anderson, M.P.; Hunt, R.D.; Griffiths, H.J.; McIntyre, K.W., Zimmerman, R.E. J. Nutr. 1977, 107, 834–839.
21. Parfitt, A.M. Metabolism 1976, 25, 809–844.
22. Malm, O.J. Scand. J. Lab. Invest. 1952, 5, 75–84.
23. Farquharson, R.F.; Salter, W.T.; Tibbets, D.M.; Aub. J. C. J. Clin. Invest.1931, 10, 221–249.
24. Bell, R.R.; Draper, H.H.; Tzeng, D.Y.M. J. Nutr. 1977, 107, 42–50.
25. Goldsmith, R.S.; Jowsey, J.; Dube, W.J.; Riggs, B.L.; Arnaud, C.D.; Kelly, P.J. J. Clin. Endocrinol. Metab. 1976, 43, 523–532.
26. Matkovic, V.; Kostial, K.; Simonovic, I.; Buzina, R.; Broderec, A., Nordin, B.E.C. Am. J. Clin. Nutr. 1979, 32, 540–549.
27. Ivey, F.J.; Shaver, K. J. Agric. Food Chem. 1977, 25, 125–130.
28. Howard, J.E.; Hopkins; T.R.; Connor, T.B. J. Clin. Endocrinol. 1953, 13, 1–19.
29. Lavender, A.R.; Pullman, T.N. Am. J. Physiol. 1963, 205, 1025–1032.
30. Amiel, C.; Kuntziger, H.; Covette, S.; Coureau, C.; Bergounioux, N. J. Clin. Invest. 1976, 57, 256–263.
31. Schuette, S.A.; Linkswiler, H.M. J. Nutr. 1982, 112, 338–349.

RECEIVED October 15, 1984

Calcium and Phosphate Needs of Preterm Infants Requiring Prolonged Intravenous Feeding

P. J. KNIGHT[1]

Departments of Surgery and Pediatrics, University of Kansas School of Medicine, Wichita, KS 67214

Premature infants who develop a gastrointestinal disease that precludes adequate oral nutrition may develop fractures and rickets when their growth is sustained by prolonged intravenous feeding. These fractures and rickets are caused by a deficiency of bone mineral substrate (calcium and/or phosphate). The goals of this chapter are: to review the previously established data regarding the amounts of calcium and phosphate normally accreted by fetuses in utero and by infants following birth, to look at the incidence of fractures and rickets with the administration of varying amounts of calcium and phosphate, to examine the physicochemical limitations of calcium and phosphate solubility in a single intravenous solution, to determine the appropriate ratio of calcium to phosphate in these solutions to maximize retention of calcium and phosphate, and to look at the clinical application of these findings.

Normal Calcium and Phosphate Accretion

Prior analyses (1-2) of the mineral content of ashed fetuses, stillborns and infant cadavers showed that the rate of calcium and phosphate accretion is highest at the end of pregnancy and that approximately half of the calcium and phosphate in the infant's body at term birth is accumulated in the last eight weeks of pregnancy - Figure 1. Consequently infants who are born prematurely start life with much lower stores of calcium and phosphate than if they had remained in the uterus for the full 40 weeks of gestation.

The enteral absorption of calcium and phosphate after birth depends on a number of factors, including the amounts ingested, the types of foods eaten and the level of vitamin D. Under optimal circumstances, the amounts of calcium and phosphate incorporated into the body per kilogram per day following birth are only about one-third of those accumulated in utero. Since the growth of the baby's body mass is faster than the absorption of calcium and phosphate, the infant must normally borrow from and redistribute his existing bone

[1]Current address: 818 N. Emporia, Suite 200, Wichita, KS 67214

0097–6156/85/0275–0041$06.00/0

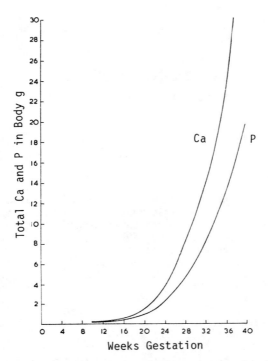

Figure 1. Total amounts and rate of accretion of calcium and phosphorus in the body of the human fetus with gestation. (Reproduced with permission from Ref. 1. Copyright 1965 W. B. Saunders Company.)

mineral stores following birth. Stated another way, the ratio of bone mineral measured as calcium relative to bone protein matrix measured as nitrogen must fall (3)-Figure 2. Since the structural strength of bone depends in part on its mineral content, eventually the mineral to osteoid ratio will fall to the point where bones break with minor trauma. Bones actively growing in the presence of a deficiency of calcium or phosphate develop excessive accumulations of noncalcified cartilage at their growth centers (epiphyseal plates), which are recognized on both physical examination and x-ray film as the deformity called rickets. Since preterm infants start extrauterine life with low bone mineral stores, these babies are particularly susceptible to developing fractures and rickets as they grow.

Incidence of Fractures and Rickets

To define the incidence of fractures and rickets that we were en- countering in infants who required prolonged parenteral feeding, we reviewed the roentgenograms of a series of preterm infants who developed necrotizing enterocolitis and who required at least four weeks of total parenteral nutritional support(4). These data are recorded in Table I.

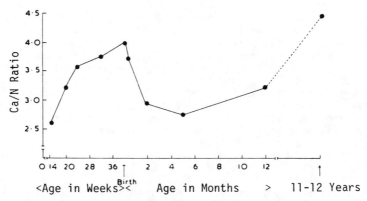

Figure 2.   Calcium-nitrogen ratio of the nonepiphyseal parts of
the human femur during development. (Reproduced with permission
from Ref. 1. Copyright 1965 W. B. Saunders Company.)

Table I.   Incidence of Demineralization, Fractures, and Rickets at
              Low and Moderate Calcium Intake

	Low Calcium Intake 1974-78 Ca = 2.5 mM PO4 = 9.7 mM	Moderate Calcium Intake 1979-80 Ca = 10 mM PO4 = 5 mM
Roentgenographically normal bones	1	0
Demineralization only	5	4
Pathologic fractures	6	1
(both rickets and fractures)	(4)	(0)
Total patients	12	5
Percent with fractures	50%	20%
Percent with rickets	33%	0%

     Infants treated before 1979 received protein hydrolysates con-
taining high concentrations of phosphate which limited the concentration of calcium that could be used without causing precipitation.
Beginning in 1979 crystalline amino acid solutions which contained
less obligatory phosphate became available; these allowed greater
latitude in the concentrations of calcium and phosphate that could be
achieved.   The data in Table I suggest that the severity of
demineralization and the incidence of fractures and rickets will
decrease if more calcium is added to the parenteral alimentation
solution.

Solubility Limitations for Ca and P

In 1978, on the basis of a few measurements of urine calcium and
phosphate excretion as well as an awareness of the previously
mentioned work regarding the amounts of calcium and phosphate
normally accreted in utero and postnatally, it became apparent that
the demineralization, fractures and rickets we were seeing in our
infants were caused by calcium deficiency. Consequently we increased the amount of calcium added to the parenteral alimentation
solutions. If more than 12.5 mM of the calcium were added to a
liter of hyperalimentation solution, gross precipitation would occur
in the feeding solution. If 10 mM of calcium were added per liter,
crystalline precipitated began to build up on the inside of our
barium-impregnated silicone rubber central venous catheters. This
crystalline precipitate resulted in gradual occlusion and functional
loss of these lines. After several false starts and six lost
catheters, chemical and crystal analysis showed that the precipitate
inside these catheters was $CaHPO_4$.
    Knowledge of some simple chemical principles and consultation
with the Handbook of Physics and Chemistry (5) resolved the problem
of the precipitates. Barium sulfate is incorporated in the silicone
rubber catheters to make them radioopaque; the Handbook shows that
$BaHPO_4$ is only slightly soluble. The barium ions in the catheter
were apparently attracting the HPO_4 ions from the hyperalimentation
solution. The concentration of phosphate at the solution-catheter
interface became high enough that $CaHPO_4$ began to precipitate out.
Propogation of the $CaHPO_4$ crystal lattice from the hyperalimentation
solution then continued until the catheter became plugged and use-
less. At slightly higher concentrations of calcium, $CaHPO_4$ would
directly precipitate out in the bags of feeding solution. The
solubility of calcium and phosphate is pH dependent as shown in
Figure 3. The pH of these feeding solutions ranges from 5.2 to 6.5
depending on the amounts of acetate, amino acids and the other
buffers in the system.
    Because of the limitation in the amount of calcium that could be
added to these high phosphate protein hydrolysates available in 1978,
we began to alternate one solution high in calcium with one solution
high in phosphate every 12 hours (6). Because of the immense
surface area of the bone crystal lattice, it was possible by
alternating solutions to get good retention of both calcium and
phosphate as determined by the measured excretion of these ions in
the urine. When the crystalline amino acid alimentation solutions
low in obligatory phosphate became available in 1979, we thought we
should determine how much calcium and phosphate could be simultaneously added to a single solution. By serially altering the concentrations of calcium and of phosphate in racks of glass test tubes
containing hyperalimentation solution, the maximal amounts that
would remain in solution were determined. These data are shown
graphically in Figure 4. The best fit curve is a hyperbola,
$x \cdot y = k$; the chemical explanation for this curve is the solubility
product for calcium dibasic phosphate, $[Ca^{++}][HPO_4^{=}] = K_{sp}$

Optimal Ca to P Ratio

Since the amounts of calcium and phosphate that can be added

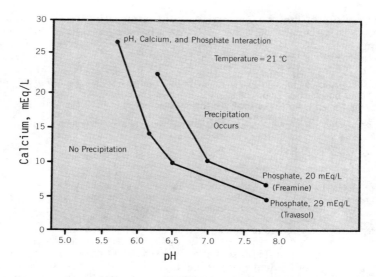

Figure 3. Interaction among pH, calcium, and phosphate in hyper-alimentation solutions. The pH, calcium values to the right of phosphate isoconcentration curves are associated with precipitation of dibasic calcium phosphate, while pH, calcium values to the left of phosphate isoconcentration curves are associated with solubility of calcium and phosphates in hyperalimentation solution. (Reproduced with permission from Ref. 6. Copyright 1980 American Medical Association.)

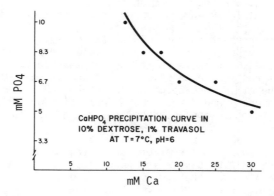

Figure 4. Precipitation became visible at the points marked by the dots. The shape of the curve (hyperbola) is described by the equation CaxP = mM^2. Solutions with CaxP products above the curve show precipitation, while solutions with CaxP products below the curve remain free of precipitate. (Reproduced with permission from Ref. 4. Copyright 1983 American Society for Parenteral and Enteral Nutrition.)

simultaneously to any parenteral alimentation solution are limited, we wanted to determine the ratio of calcium to phosphate that would maximize retention of these elements by the body. We had a ballpark estimate for this optimal Ca/P ratio from several pieces of data. In adults, bone is the repository for 98% of the body's calcium and 85% of the body's phosphate; in infants, the skeleton contains about 96% of the body's calcium and 70% of its phosphate. The Ca/P ratio of bone mineral (hydroxyapatite or $3[Ca_3(PO_4)]\cdot Ca(OH)_2$ can be expressed in various units yielding different values:

	Moles	Equivalents	Weight
$\dfrac{Ca}{P}$	$\dfrac{10}{6} = 1.67$	$\dfrac{20}{18} = 1.11$	$\dfrac{10 \times 40}{6 \times 31} = 2.15$

The remaining non-osseous body calcium is largely dissolved in body fluids; the ionized portion of this soluble calcium mediates muscle contraction including life-sustaining cardiac muscle and respiratory muscle contraction. Acute serum calcium deficiencies rarely develop because the body's homeostatic mechanisms can mobilize calcium from bone to maintain the serum calcium level until none is left in bone. The remaining non-osseous phosphate is: structurally incorporated in the phospholipids of cell membranes and in the nucleic acids, is the principle intracellular buffer of the body, and is the primary mediator of intracellular energy transfer. Acute phosphate deficiencies do occur; they can present as cardiac (7) or respiratory (8) or peripheral muscle failure (9) or as infections due to the inability of polymorphonuclear leukocytes to phagocytize (10). Since the effects of phosphate deficiency can occur acutely and are life-threatening as opposed to the fractures and rickets of chronic body calcium depletion, one should err on the side of giving more phosphate than calcium. Any excess phosphate is excreted in the urine.

Our simple approach to determining the optimal Ca/P ratio for intravenous feeding solutions was to simply alter the ratio of calcium to phosphate in these solutions and measure the only external loss of calcium and phosphate which was in the urine. We initially assumed that the difference between the intake and urinary loss of calcium and phosphate would measure the retention of these elements. The results of seven balance studies at varying Ca/P ratios are shown in Figure 5.

The observed calcium/phosphate ratio of 4.5 at the intercept of the calcium and phosphate retention curves that should minimize the sum of the urine calcium plus urine phosphate losses was difficult to believe in view of both the known Ca/P ratio of bone and the amounts we were adding to these solutions. This disparity between the optimal ratio determined experimentally and what we had assumed this ratio should be on the basis of known body composition is partially reconciled by the experiment of Sutton and Barltrop. They fed preterm infants stable Ca[46] and observed that up to 20% of the isotope absorbed was subsequently excreted in the stool. Our infants also were undoubtedly having unmeasured calcium losses from the bile, pancreatic juice and succus entericus secreted into their intestine

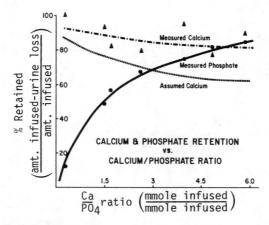

Figure 5.   The areas above the measured calcium and measured
phosphate retention curves represent the percent of the intra-
venously administered calcium and phosphate that was lost in the
urine.   The combined percent losses (Ca + P) are minimized at the
intercept of the curves.   The dotted line represents an assumed
endogenous fecal loss of 20% of the infused calcium added to the
measured urinary calcium losses.   The Ca/P ratio that minimizes
the percent calcium and phosphate losses is then approximately
3.0.   (Reproduced with permission from Ref. 4.   Copyright 1983
American Society for Parenteral and Enteral Nutrition.)

and this calcium was not reabsorbed. If one assumes that 20% of
absorbed calcium is subsequently excreted into the bowel, the
calcium retention curve would move downward, resulting in an intercept with the phosphate retention curve at a Ca/P ratio of about
3.0-Figure 5. Although this Ca/P ratio still seems a bit high, Ca/P
ratios between 2.0 and 3.0 can be maintained for prolonged periods
without calcium or phosphate depletion occurring and with the urine
excretion both of these elements being low but measurable.

Clinical Application

Using the in vitro determined maximal Ca x P product (mM^2) of 75 for
barium-impregnated silicone rubber catheters and of 100 for stainless
steel needles and polyethylene catheters and the Ca/P ratio determined by the interception of the retention curves at a value of 2.0
to 3.0, one has two equations with two unknowns. Solving these
equations simultaneously gives the average optimal concentrations
for calcium and phosphate in the alimentation solutions. These are
concentrations that will both stay in solution and will maximize
retention for the average patient. Although these average concentrations are useful starting points, one would like to be able to
fine-tune these concentrations to each individual patient's metabolism. This metabolism is determined by the patient's parathormone
and calcitonin levels, the active vitamin D level, the body's acid-
base status and other poorly understood factors. These factors
acting on the renal tubular cells determine the percent of the
calcium and phosphate filtered by the glomeruli that is reabsorbed.
In the absence of renal tubular defects, measuring the concentrations
of calcium and phosphate excreted in any individual infant's urine
indicated in which direction the Ca/P ratio should be changed to
achieve this fine-tuning. When the Ca/P ratio is appropriate, both
calcium and phosphate should be measurable; absence of measurable
calcium or phosphate in the urine means the ratio must be changed.
Our data from premature infants indicate that if both the urine
calcium and the urine phosphate are between 2 and 10 mg/dl, the ratio
of these elements in the alimentation solution is appropriate.

A Lesson from the Body

The product of the Ca x P concentrations maintained in human serum
has a ten-fold safety factor over the amounts that cause precipitation. Some of this safety factor is because only slightly more
than half of the serum calcium is actively available (ionized
calcium), while the remainder is either bound to protein or chelated
by cirtate ion. If citrate buffer is added to the alimentation
solutions to a concentration of 5 mM, the Ca x P product can excede
1800 without precipitation of $CaHPO_4$ occurring. This effect of
citrate is not solely due to chelation since 5 mM of citrate can only
bind 7.5 mM of calcium. The citrate ions appear to prevent $CaHPO_4$
crystals from propagating and coming out of solution as shown by
Patterson (12). Since recognizing the Ca x P products and the Ca/P
ratio which are appropriate for preterm infants on prolonged intravenous alimentation, we have had no further cases of fractures or
rickets in our preterm infants on prolonged intravenous nutrition.

Literature Cited

1. Widdowson, E. M.; McNance, R. A. Pediatr. Clin. of N. Amer. 1965, 12, 595-614.
2. Shaw, J. C. L. Pediatr. Clin. of N. Amer. 1973, 20, 333-58.
3. Dickerson, J. W. T. Biochem. J. 1962, 82, 56-61.
4. Knight, P. J.; Heer, D.; Abdenour, G. J. E. P. N. 1983, 7, 110-14.
5. Weast, R. "Handbook of Physics and Chemistry"; Cleveland, C. R. C. Press, 1974; p. B 71 and B 78.
6. Knight, P. J.; Buchanan, S.; Clatworthy, H. W. J. A. M. A. 1980, 243, 1244-46.
7. O'Connor, L. R.; Wheeler, W. S.; Bethune, J. E. New Engl. J. Med. 1977, 297, 901-03.
8. Newman, J. H.; Neff, T. A.; Ziporin, P. New Engl. J. Med. 1977, 296, 1101-03.
9. Weintraub, M. I. J. A. M. A. 1976, 235, 1040-41.
10. Craddock, P. R.; Yawata, Y.; VanSaten, L.; et.al. New Engl. J. Med. 1974, 290, 1403-07.
11. Sutton, A.; Barltrop, D. Nature 1973, 242, 265.
12. Patterson, D. Nature 1954, 173, 75-76.

RECEIVED December 26, 1984

Dietary Phytate and Calcium Bioavailability

ERNST GRAF[1] and JOHN W. EATON[2]

[1] The Pillsbury Company, Minneapolis, MN 55414
[2] Department of Laboratory Medicine and Pathology, University of Minnesota, Minneapolis, MN 55455

Phytic acid (myo-inositol hexaphosphoric acid) comprises 1 to 5% by weight of legumes, cereals, oil seeds and nuts and thus is present in numerous food commodities. This compound has been indicted as a major cause of several mineral deficiency syndromes, through the formation of insoluble metallophytate complexes. Our recent results demonstrate, however, that polyvalent cations form soluble complexes with phytate at high phytate to metal ratios and that phytate does not impair the absorption of $^{45}Ca^{2+}$ administered to mice by gavage. Thus, many of the supposedly adverse effects of dietary phytate and the resulting mineral deficiencies may merely reflect general dietary inadequacy. Furthermore, some dietary phytate may actually be beneficial due to its ability to suppress iron-mediated oxidative processes. Indeed, phytic acid may reduce the incidence of colonic cancer and function as a natural food preservative.

Calcium is the most abundant mineral in the human body. About 99% is deposited as phosphate matrices in bones and teeth. In addition to this structural role, Ca^{2+} is of central importance to most aspects of cell physiology, both as a cofactor for extracellular enzymes and as an intracellular regulator. Intracellular Ca^{2+} has profound effects on a variety of cellular and enzyme functions, including excitation-contraction coupling in all forms of muscle, excitation-secretion coupling at nerve endings and in both exocrine and endocrine glands, exocytosis, membrane transport, cell morphology, gene expression, and vision.

Recognizing the importance of adequate Ca^{2+} intake,

0097-6156/85/0275-0051$06.00/0

the United States National Research Council recommends an intake of 800 mg Ca^{2+} (recommended daily allowance or RDA) and 1200 mg during pregnancy and lactation. In order to meet these guidelines, certain food staples in the U.S. are fortified with Ca^{2+}. Despite this widespread effort, Ca^{2+} deficiency is still very common as evidenced by the high incidence of osteoporosis. This health problem partially arises from the fact that maintenance of Ca^{2+} homeostasis requires not only adequate dietary intake, but intestinal absorption and subsequent utilization of Ca^{2+}.

Bioavailability of Ca^{2+} is affected by numerous physiological conditions, including age, sex, genetic make-up, stress, hormonal status, health status, and nutritional habits. In addition to these intrinsic factors, certain dietary components, such as fiber and oxalate, form insoluble complexes with Ca^{2+} and interfere with its absorption. Another putative culprit in this category is phytic acid.

Phytic acid (Figure 1) constitutes 1 to 5% by weight of most nutritionally important plant seeds (Table I) and

Table I. Phytic Acid Content of Selected Seeds

	Phytic Acid (% w/w)	Reference
Millet	0.7	2
Barley	0.7	2
Sorghum	0.9	2
Corn	0.9	2
Rye	0.9	2
Dehydrated peas	0.9	3
Oat	1.0	2
Wheat	1.1	2
Soybeans	1.4	4
Sunflower seeds	1.9	5
Peanuts	1.9	3
Wild rice	2.2	5
Lima beans	2.5	5
Sesame seeds	5.3	3

typically accounts for 60 to 90% of the total phosphorus. It usually occurs in discrete regions of the seeds, such as the aleurone layer of wheat and rice (6), where it is believed to serve as an antioxidant and protect against oxidative damage during storage (7). Due to the ability of phytic acid to precipitate polycationic nutrilites (8,9), its presence in food has concerned nutritionists for several decades. Several recent reviews extensively discuss the chemistry and nutritional ramifications of dietary phytate (4,10-15).

The effect of phytic acid on Ca^{2+} bioavailability is still in dispute. Some early nutritionists reported rachitogenic properties of dietary phytate based on feeding studies using puppies (16-18) and epidemiological studies on British-born children of Asian immigrants

(19,20). These conclusions were seriously questioned (21-24) and recent results indicate that the bioavailability of Ca^{2+} is the same from a casein diet as from a high-phytate soy concentrate (25). Oberleas contends that, in the presence of adequate amounts of Ca^{2+} and vitamin D, dietary phytate is not rachitogenic, even though it may bind substantial amounts of Ca^{2+} (15). This controversy, the relative paucity of available information, and the growing incidence of Ca^{2+} deficiency prompted us to investigate further the chemical interactions between Ca^{2+} and phytate and to assess its effect on the bioavailability of Ca^{2+} administered to mice by gavage.

Methods

Phytic acid solutions were prepared by titrating sodium phytate (Sigma Chemical Company) with HCl; the concentration was determined by analyzing for inorganic phosphate after wet ashing with H_2SO_4-HNO_3 (3:2) for 45 minutes. The concentration of $CaCl_2$ stock solutions was measured by atomic absorption spectrometry.

Ca^{2+} solubility was determined by incubating Ca^{2+} and phytate in 100 mM HEPES pH 7.4 at 25°C for 2 hours, centrifuging and measuring $^{45}Ca^{2+}$ in the supernatant by liquid scintillation counting (26).

The binding of Ca^{2+} to phytic acid was ascertained by measuring Ca^{2+} potentiometrically with a Radiometer Ca^{2+}-selective electrode (27).

For the quantitation of *in vivo* absorption of $^{45}Ca^{2+}$, 200 μl of radioactive solutions were administered by gastric gavage to male mice kept on a deionized water diet during the previous 18 hours. After 4 hours blood was obtained by axillary incision and $^{45}Ca^{2+}$ was extracted with TCA and determined by liquid scintillation counting (26).

Iron-mediated generation of hydroxyl radical (·OH) was monitored by the hypoxanthine-xanthine oxidase method as previously described (28). Formaldehyde produced by reaction of ·OH with DMSO was determined spectrophotometrically by the Hantzsch reaction (29).

Lipid peroxidation was measured by determining malondialdehyde (MDA) spectrophotometrically by the thiobarbituric acid method (30).

Results and Discussion

Traditionally, polyvalent cation-phytate complexes have been considered uniformly insoluble, and this dogma has been the basis of most methods for the determination of phytate (31). In 1976, however, soluble monoferric phytate was isolated (32), and most recently, Ca_1- and Ca_2-phytate were also found to be soluble (27). Figure 2 shows a Scatchard plot of Ca^{2+} binding to phytate at pH 4.8. At a ratio of 2.2 to 2.4 the complex precipitates. This ratio is independent of the phytate concentration over a 100-

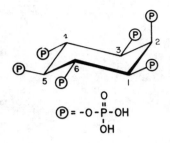

Figure 1. Structure of phytic acid in dilute solution (<u>1</u>).

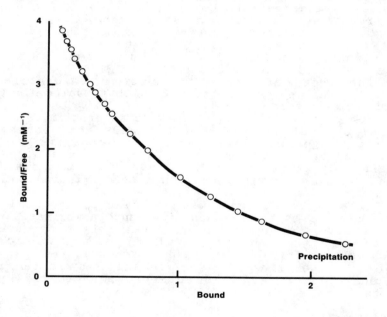

Figure 2. Scatchard plot of binding of Ca^{2+} to phytic acid. A solution of 3.04 mM phytic acid in 50 mM acetic acid pH 4.8 was titrated with 1.005 M $CaCl_2$ at 40^oC.
(Reproduced from Ref. 27.   Copyright 1983 American Chemical Society.)

fold concentration range, suggesting that Ca_1-phytate and possibly Ca_2-phytate are two very soluble species. Qualitatively similar results were observed at higher pH, yet a quantitative analysis is rendered extremely difficult, as the high affinity constants of phytate for Ca^{2+} at higher pH lower the free Ca^{2+} concentration below the detection limit of Ca^{2+}-selective potentiometry (27). The high solubility of Ca^{2+}-phytate at pH 7.4 is further illustrated in Figure 3. The addition of small amounts of phytate to a solution of 10 mM Ca^{2+} forms a Ca_6-phytate precipitate. However, with increasing phytate, soluble complexes of lower Ca^{2+} to phytate ratios are formed. At the highest phytate concentration, Ca^{2+} is completely soluble in the form of Ca_1-, $Ca_{1/2}$- and possibly Ca_2-phytate. We observed similar solubility behavior of phytate complexes with Ba^{2+}, Cd^{2+}, Co^{2+}, Cu^{2+}, Fe^{2+}, Hg^{2+}, Mg^{2+}, Mn^{2+}, Ni^{2+}, Pb^{2+}, Sn^{2+}, Sr^{2+}, Zn^{2+}, Al^{3+}, Fe^{3+}, and La^{3+}, which suggests that all polyvalent cation phytate chelates of equimolar stoichiometry are soluble, in conflict with existing views on metallophytate properties.

The trough in Figure 3 occurs at Ca_6-phytate, the least soluble species. This throws into question the validity of the widely accepted dogma that increasing the metal to phytate ratio improves metal bioavailability. Obviously, increasing the Ca^{2+} concentration on the right side of the trough in Figure 3 will not significantly alter the concentration of soluble Ca^{2+}, but will greatly reduce the solubility of trace metals, such as Zn^{2+} and Fe^{3+}, as they are bound to phytate much more tightly than Ca^{2+} (26). Similarly, reducing the phytate concentration will only reduce the amount of soluble Ca^{2+} and other polycationic nutrilites.

Ca^{2+} solubility is lowest at a Ca^{2+} to phytate ratio of 6:1. Nevertheless, Ca_6-phytate is not completely insoluble as shown in Figure 4. Half-maximal solubility occurs at 0.70 mM $Ca(phy)_{1/6}$. Below 0.1 mM, the complex is completely soluble. At high concentrations, the absolute amount of soluble Ca^{2+} increases, despite decreasing %-solubility. All six dissociation constants for Ca_n-phytate complexes are much lower than their respective solubility products, which permits the existence of soluble Ca_n-phytate complexes for n of any value between 1 and 6. Figure 4 represents the most adverse condition for Ca^{2+} solubility, and any decrease or increase in phytate, Ca^{2+}, or both at any ratio would raise the soluble Ca^{2+} concentration.

Without considering the complexity of additional interactions of phytate and Ca^{2+} with other polyvalent cations, proteins, and fiber, Figures 3 and 4 predict that the progressive elimination of bran in Table II lowers the concentration of soluble Ca^{2+}. This effect is especially pronounced in the bread, where part of the remaining phytate was hydrolyzed by yeast phytase during leavening (Table II). These conclusions and their implications

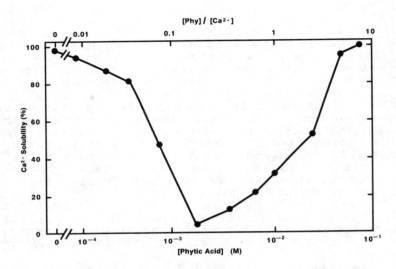

Figure 3. Effect of phytic acid on the solubility of 10 mM Ca^{2+} in 100 mM HEPES pH 7.4. (Reproduced with permission from Ref. 26. Copyright 1984 J. Nutr., American Institute of Nutrition.)

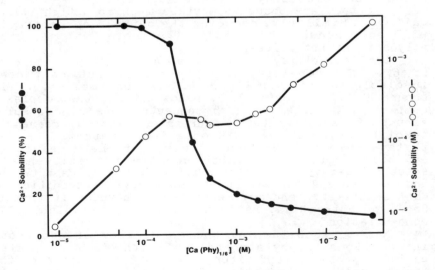

Figure 4. Solubility of Ca_6-phytate in 100 mM HEPES pH 7.4.

Table II. Ca^{2+} and Phytate Contents of Wheat Products (<u>2</u>)

	Phytate ($\%$ w/w)	Calcium ($\%$ w/w)	Phytate:Ca^{2+} (mol/mol)
Wheat	1.1	0.120	0.6
Flour (100% extraction rate)	0.96	0.035	1.7
Flour (85% extraction rate)	0.52	0.020	1.6
Flour (70% extraction rate)	0.38	0.015	1.6
Bread (100% extraction rate)	0.66	0.035	1.1
Bread (85% extraction rate)	0.25	0.020	0.8
Bread (70% extraction rate)	0.14	0.015	0.6

concerning Ca^{2+} bioavailability are an oversimplification, due to additional Ca^{2+} and other polyvalent cations present in the remaining diet and due to the question of bioavailability of soluble Ca^{2+}-phytate. However, our experimental results do provide a viable explanation for the previous observation that iron from mature soybeans is more bioavailable than that from immature seeds despite the 3-fold higher phytate concentration in the former (<u>33</u>). Similarly, Reinhold et al. (<u>23</u>,<u>24</u>) have shown that tanok and bazari - two Iranian breads of the same fiber content but greatly different phytate content - inhibited calcium and zinc uptake equally.

Phytate has no detrimental effects on the bioavailability of Ca^{2+} administered to mice by gastric gavage (Figure 5). Soluble Ca_1-phytate (ratio = 10), soluble Ca_4-phytate (ratio = 0.25, 0.1 mM_2 Ca^{2+}), precipitated Ca_4-phytate (ratio = 0.25, 1.0 mM Ca^{2+}), and a mixture of various species (ratio = 1) were absorbed to the same extent as unchelated Ca^{2+}. This seems surprising, considering that the total uptake of Ca^{2+} was directly proportional to the total Ca^{2+} administered as the absorption of Ca^{2+} at 0.1 mM ranged between 0.116 and 0.164 nmoles and at 1.0 mM it ranged between 1.48 and 1.70 nmoles Ca^{2+} absorbed. Under normal physiological conditions these absorption values may change slightly. We have previously shown that the absorption of iron administered to mice by duodenal instillation is also unaffected by phytic acid (<u>26</u>). These results are in conflict with some prior animal feeding studies, most of which were conducted over longer periods of time using complex diets that also included fiber, another mineral-binding component. The multitude of intrinsic and extrinsic factors influencing intestinal absorption of minerals may account for the numerous controversial reports in the literature, and caution must be exercised in generalizing effects observed under a somewhat unphysiological experimental condition.

From our results on solubility properties of Ca^{2+}-phytate complexes and their bioavailability to mice we conclude that, at least in our experimental system, phytate has no detrimental effect on Ca^{2+} bioavailability. Indeed, as we have argued elsewhere, adequate levels of dietary phytate may actually be beneficial due to the ability of phytate to suppress the formation of hydroxyl radical ($\cdot$OH) and lipid peroxidation, two oxidative processes partly responsible for rancidity and spoilage of foods and for several diseases ($\underline{7},\underline{28},\underline{30}$).

The catalysis by iron of the generation of $\cdot$OH, a highly reactive and dangerous species of activated oxygen, has been well established:

$$O_2^{\overline{\cdot}} + Fe^{3+} \longrightarrow Fe^{2+} + O_2$$

$$2O_2^{\overline{\cdot}} + 2H^+ \longrightarrow H_2O_2 + O_2$$

$$H_2O_2 + Fe^{2+} \longrightarrow Fe^{3+} + OH^- + \cdot OH$$

In this classical Haber-Weiss cycle iron is being reduced by superoxide anion radical ($O_2^{\overline{\cdot}}$), ascorbic acid or glutathione and subsequently decomposes hydrogen peroxide - formed by spontaneous dismutation of $O_2^{\overline{\cdot}}$ - in the Fenton reaction to produce $\cdot$OH. This iron-driven $\cdot$OH formation has a stringent requirement for an available iron coordination site, a <u>sine qua non</u> met not only by hexaaquoiron(III) but by most iron chelates ($\underline{28}$). Thus, Fe-EDTA, -EGTA, and -ATP retain a reactive coordination site and catalyze the Haber-Weiss cycle. Phytic acid, however, occupies all available iron coordination sites and consequently fails to support $\cdot$OH generation (Figure 6).

The occupation of all available coordination sites by phytate suppresses other iron-mediated processes, such as lipid peroxidation ($\underline{30}$). Figure 6 demonstrates that 0.24 mM phytate prevents the peroxidation of arachidonic acid driven by ascorbic acid and iron, whereas substantial amounts of malondialdehyde arise in the presence of free iron or of an iron-ADP chelate.

The obliteration of these iron-dependent oxidative processes by phytate suggests that this ubiquitous and abundant plant component functions as a natural antioxidant, preventing oxidative damage during storage of seeds. Surface treatment with phytic acid of various fruits and vegetables preserved their color, texture and flavor ($\underline{7}$). Thus, dietary phytate may be a superior substitute for presently employed food preservatives, many of which pose significant health hazards. Additional applications of phytic acid are summarized in a recent review ($\underline{34}$).

By virtue of rendering iron catalytically inactive, dietary phytate may also suppress the incidence of colonic cancer ($\underline{35}$). Intestinal aerobic bacteria and/or minor inflammatory events generate substantial amounts of $O_2^{\overline{\cdot}}$ leading to $\cdot$OH formation and lipid peroxidation. These two processes are thought to be important elements in tissue injury which occurs during inflammation. This argument is compatible with the observation that colonic cancer is frequently preceded, or accompanied, by pigmentation of the colonic epithelium; lipofuscin, a byproduct of lipid

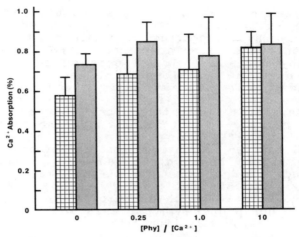

Figure 5. Ca^{2+} bioavailability to mice. The animals were administered by gavage 200 μl of 0.1 mM (cross-hatched) and 1.0 mM $CaCl_2$ (dotted) containing variable amounts of phytic acid as indicated on the abscissa. The mean and standard deviation of 4 to 9 determinations of Ca^{2+} absorption into the blood after 4 hours are shown on the ordinate. (Reproduced with permission from Ref. 26. Copyright 1984 J. Nutr., American Institute of Nutrition.)

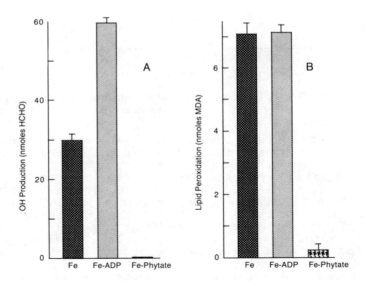

Figure 6. Effect of ADP and phytate on iron-mediated •OH formation (A) and lipid peroxidation (B). The mean and standard deviation of 3 determinations are shown.

peroxidation, mainly constitutes the pigment granules (36). The suppression of iron-catalyzed oxidative processes by phytate provides a viable biochemical basis for the epidemiological inverse correlation between dietary phytate and the incidence of colonic cancer. Figure 7 shows the relationship between consumption of cereals and colonic cancer rates in 37 countries (37). The high correlation coefficient of -0.30 appears to also support Burkitt's contention, that dietary fiber is inversely related to the colonic carcinogenicity of a particular diet (38). On closer examination of this hypothesis, however, we noticed that of several classes of food with high fiber content (cereals, starch foods, vegetables, and fruits), only the intake of cereals - containing 1 to 5% phytic acid - shows the expected negative correlation with colonic cancer (37). Further support for our proposed hypothesis derives from other epidemiological investigations (35).

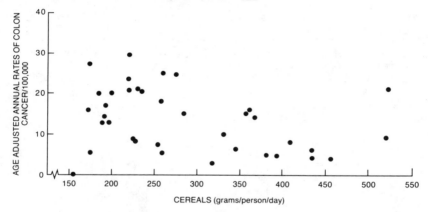

Figure 7. Correlation between colon cancer rate and the intake of cereals in 37 countries. (Adapted from Ref. 37.)

    In conclusion, phytic acid forms soluble complexes with Ca^{2+} at intestinal pH under a variety of conditions and fails to inhibit Ca^{2+} bioavailability to mice in our experimental system. Despite the hazard in direct extrapolation of results obtained with animals kept on a well-defined dietary regimen to humans consuming a complex diet, many elements of which affect Ca^{2+} bioavailability, our data demonstrate the need for a reevaluation of the putative antinutritional properties of dietary phytate. Our further contention that adequate levels of dietary phytate may actually be beneficial due to its food preserving properties and its protection against colonic cancer will warrant a prospective epidemiological human study designed to assess the longterm effects of dietary phytate on mineral bioavailability and inflammatory bowel diseases.

Literature Cited

1. Johnson, L.F.; Tate, M.E. Can. J. Chem. 1969, 47, 63.
2. Lepen, B.; Adrian, J. Industrie des céréales 1982, 16, 3.
3. Graf, E.; Dintzis, F.R. J. Agric. Food Chem. 1982, 30, 1094.
4. Erdman, J.W. J. Am. Oil Chem. Soc. 1979, 56, 736.
5. Harland, B.F.; Prosky, L. Cereal Foods World 1979, 24, 387.
6. Tanaka, K.; Yoshida, T.; Kasai, Z. Plant Cell Physiol. 1974, 15, 147.
7. Graf, E.; Mahoney, J.R.; Eaton, J.W. Manuscript submitted.
8. Maddaiah, V.T.; Kurnick, A.A.; Reid, B.L. Proc. Soc. Exp. Biol. Med. 1964, 115, 391.
9. Vohra, P.; Gray, G.A.; Kratzer, F.H. Proc. Soc. Exp. Biol. Med. 1965, 120, 447.
10. Maga, J.A. J. Agric. Food Chem. 1982, 30, 1.
11. Cheryan, M. CRC Crit. Rev. Food Sci. Nutr. 1980, 13, 297.
12. Cosgrove, D.J. "Inositol Phosphates: Their Chemistry, Biochemistry, and Physiology"; Elsevier Scientific Publishing Company: New York, 1980.
13. O'Dell, B.L. In "Soy Protein and Human Nutrition"; Wilcke, H.L.; Hopkins, D.T.; Waggle, D.H., Eds.; Academic Press: New York, 1979; pp. 187-207.
14. Reddy, N.R. Adv. Food Res. 1982, 28, 1.
15. Oberleas, D. In "Toxicants Occurring Naturally in Foods"; Committee on Food Protection, Food and Nutrition Board, National Research Council, Ed.; Natl. Acad. Sci.: Washington, D.C., 1973; pp. 187-207.
16. Mellanby, E. Spec. Dept. Ser. Med. Res. Council, London 1925, No. 93.
17. Bruce, H.M.; Callow, R.K. Biochem. J. 1934, 28, 517.
18. Mellanby, E. J. Physiol. 1949, 109, 488.
19. Wills, M.R.; Day, R.C.; Phillips, J.B.; Bateman, E.C. Lancet 1972, i, 771.
20. Goel, K.M.; Logan, R.W.; Arneil, G.C.; Sweet, E.M.; Warren, J.M.; Shanks, R.A. Lancet 1976, i, 1141.
21. Walker, A.R.P.; Fox, W.F.; Irving, J.T. Biochem. J. 1948, 42, 452.
22. Walker, A.R.P. Lancet 1951, 261, 244.
23. Reinhold, J.G. Lancet 1976, ii, 1132.
24. Reinhold, J.G.; Faradji, B.; Abadi, P.; Ismail-Beigi, F. J. Nutr. 1976, 106, 493.
25. Forbes, R.M.; Weingartner, K.E.; Parker, H.M.; Bell, R.R.; Erdman, J.W. J. Nutr. 1979, 1109, 1652.
26. Graf, E.; Eaton, J.W. J. Nutr. 1984, 114, 1192.
27. Graf, E. J. Agric. Food Chem. 1983, 31, 851.
28. Graf, E.; Mahoney, J.R.; Bryant, R.G.; Eaton, J.W. J. Biol. Chem. 1984, 259, 3620.
29. Nash, T. Biochem. J. 1953, 55, 416.
30. Graf, E.; Sadrzadeh, S.M.H.; Eaton, J.W. Manuscript submitted.

31. Oberleas, D. Meth. Anal. Biochem. 1971, 20, 87.
32. Morris, E.R.; Ellis, R. J. Nutr. 1976, 106, 753.
33. Welch, R.M.; VanCampen, D.R. J. Nutr. 1975, 105, 253.
34. Graf, E. J. Am. Oil Chem. Soc. 1983, 60, 1861.
35. Graf, E.; Eaton, J.W. Manuscript submitted.
36. Ghadially, F.N.; Parry, E.W. J. Path. Bact. 1966, 92,
    313.
37. Irving, D.; Drasar, B.S. Br. J. Cancer 1973, 28, 462.
38. Burkitt, D.P. Cancer 1971, 1, 3.

RECEIVED October 15, 1984

Bioavailability of Dietary Calcium
Effect of Phytate on Adult Men Consuming Nonvegetarian Diets

EUGENE R. MORRIS and REX ELLIS

Beltsville Human Nutrition Research Center, Agricultural Research Service, U.S. Department of Agriculture, Beltsville, MD 20705

Two metabolic balance studies were conducted using healthy adult men to study the effect of phytate on bioavailability of dietary calcium. Dietary treatments were each 15 days in duration. In the first study, a mean daily calcium balance of 208 ± 58 (SD) mg was observed when 2.0 g of phytate from 36 g of whole wheat bran was consumed daily with 1100 mg of calcium, phytate/calcium molar ratio 0.11. Calcium balance was 184 ± 87 mg when 36 g of dephytinized bran was consumed with the same intake of calcium, phytate/calcium molar ratio 0.01. In the second study, calcium intake was 740 mg/day. Apparent absorption (intake minus fecal excretion) of calcium decreased when the diet contained muffins with added sodium phytate to increase the molar ratio of phytate/calcium from 0.04 to 0.14 and 0.24. One-half of the men excreted more calcium in feces than was consumed when the high phytate diet was consumed. People consuming diets with molar ratios of phytate/calcium exceeding 0.2 may be at risk of calcium deficiency because of low bioavailability of dietary calcium unless physiological adjustments can be accomplished that maintain homeostasis.

Phytate from plant products is generally considered to decrease bioavailability of dietary calcium to humans (1). Most human studies investigating the effect of phytate on calcium bioavailability have used wheat bran or high extraction wheat flour as the source of phytate. Reinhold et al. (2) observed greater negative calcium balances by adult men consuming unleavened whole meal bread compared to the same amount of phytate as sodium phytate in a low fiber bread. With the possibility therefore, that there may be an effect of dietary fiber on calcium bioavailability, studies are needed to clarify whether the phytate or dietary fiber component of wheat bran is the responsible agent decreasing calcium bioavailability.

Two metabolic balance studies conducted in our laboratory have yielded information relative to the effect of phytate and dietary fiber on calcium bioavailabilty. In the first study, a relatively high intake of dietary fiber was consumed with a 10-fold difference in phytate intake from wheat bran. In the second study three levels of phytate were consumed with a low amount of dephytinized bran as the principal dietary fiber source. The two higher phytate levels in the latter study were attained using sodium phytate.

Materials and Methods

Balance study protocols

The two human diet studies will be designated HS-I and HS-II. The first study (HS-I) was 30 days in duration. Ten adult men in good health ranging in age from 23 to 48 years were subjects. The menus consisted of foods routinely consumed in the United States and repeated in 5-day cycles (3).
     The diet treatments were level of phytate intake, either 0.2 or 2.0 g/day. Each level was consumed for 15 days, three consecutive repeats of the 5-day menu cycle. To provide 2.0 g/day of phytic acid, 36 g of wheat bran was baked into 6 muffins and two muffins were eaten each meal. Dephytinized bran was prepared by incubating the bran in water and allowing the endogenous phytase to hydrolyze the phytate, then the entire incubation mixture was freeze-dried (4) and 36 g baked into 6 muffins. Thus, the intake of all nutrients and neutral detergent fiber was the same for both phytate intakes. Five subjects consumed the whole bran muffins for 15 days followed by the dephytinized bran muffins for 15 days and the other 5 subjects in the reverse order. Brilliant blue dye was given at breakfast on the first day of each collection period to aid in demarcation of stools. Stool composites were made for days 1-5, 6-15, 16-20 and 21-30 and urine composites for days 6-15, and 21-30. Daily food composites were made, homogenized, freeze-dried and then analyzed to determine mineral nutrient intakes.
     Twelve adult men participated in the second study (HS-II). The diet treatments consisted of three different phytate intakes. A three period crossover design was used in which each sequence was repeated the same number of times (Table I). A repeating 5-day menu cycle was again used and each dietary treatment was 15 days in duration, total time for the study was 45 days. Muffins, two each meal, contained 8.65 g of the water insoluble fraction of dephytinized bran per day, derived from 12 g of whole wheat bran. The muffin ingredients, including the dephytinized bran, provided about 0.3 g of phytate per day. Sodium phytate (Sigma Chemical Co., St. Louis, MO) was added to the muffins to provide two higher intakes of phytate. The other menu items contributed about 0.2 g of phytate per day. Brilliant blue dye was again administered to aid in demarcation of stools. Urine and feces were collected throughout the entire study and composites were made from each 5 days.

Table I.   Study Design for HS-II

Study Days	Diet Treatment Assignment (12 subjects)					
	(2)	(2)	(2)	(2)	(2)	(2)
1-15	A	A	B	B	C	C
16-30	B	C	A	C	A	B
31-45	C	B	C	A	B	A

Note: Menus repeated in 5-day cycles. Diet treatments were phytic acid intake, g/day, A, 0.5; B, 1.7; and C, 2.9.

Concurrently with HS-II, 6 additional men consumed the low phytate menus for 45 days, 9 consecutive repeats of the 5-day menu cycle. This study will be designated HS-III in the following section. The non-muffin food items were provided in increments as needed to maintain volunteer's weight $\pm$ 2 kg during each of the three studies.

For one week prior to and following the 45-day controlled diet period of HS-II and HS-III, each individual collected duplicate portions of all food and liquids consumed. These self-selected diets were brought to the laboratory and daily composites were prepared for analysis.

The wheat bran used in these studies was milled for us from a single lot of Waldron hard red spring wheat. Other foods and diet ingredients were purchased from local food suppliers. Data from HS-I was analyzed statistically by Student's paired t test, each subject acting as his own control. A three-way analysis of variance (ANOVA) was performed to test for significant differences betwen diet treatments, periods and individuals in HS-II and HS-III.

Phytate analysis of muffins, diet composites and stools was performed by a modification of the ferric ion precipitation method for HS-I (5) and by an ion exchange method for HS-II and -III (6). Calcium was determined by atomic absorption spectrophotometry. HS-I diet and fecal samples were dry ashed and HS-II and -III samples were prepared by a combination dry-wet ashing procedure (Hill et al. unpublished). The approved method of American Association of Cereal Chemists (7) was used for determining neutral detergent fiber.

Results

Dietary intake data for calcium, phytate and phytate/calcium ratio are summarized in Table II. Mean calcium intake for HS-I was about 300 mg greater than the recommended dietary allowance (RDA) of 800 mg established for adults by the National Research Council (8) and for HS-II just slightly less than the RDA. The range of calcium intakes, because of different caloric needs was from 927 to 1490

Table II.   Summary of Dietary Intakes

Calcium	Phytate [1]	Phytate/Calcium
mg/day	g/day	molar ratio
	HS-I[2]	
1090	2.0	0.11
1150	0.2	0.01
	HS-II[3]	
740	0.5	0.04
740	1.7	0.14
740	2.9	0.24

[1]As phytic acid.
[2]Ten subjects, means for 15 days.
[3]Twelve subjects, means for 15 days. Intakes of HS-III
(6 subjects) same as low phytate diet.

for HS-I and 620 to 910 for HS-II. Phytate intakes were essentially constant across individuals because almost all of the phytic
acid was present in the muffins and each individual ate two muffins
each meal regardless of caloric needs. In HS-I there was a 10-fold
difference in the two phytate intakes and the resultant phytate to
calcium molar ratios. The highest phytate intake during HS-II was
only 0.9 g/day greater than HS-I, but the intermediate and high
phytate/calcium molar ratios were greater than for HS-I.
    HS-I. Metabolic balance data for HS-I are summarized in Table
III. For the first 5 days of each diet treatment, A in Table III,
about 100 mg more of calcium was excreted in the feces when whole
bran muffins were consumed compared to dephytinized bran muffins.
Consequently, apparent absorption was greatest, $P < 0.05$, when the
phytate intake was low i.e., when dephytinized bran muffins were
consumed. For the subsequent 10 days, days 6-15, fecal excretion
was almost the same for each diet treatment, consequently, no
difference in apparent absorption. During the same period, urinary
calcium excretion was slightly lower when whole bran muffins were
consumed and balance was almost 25 mg greater than when eating
the dephytinized bran muffins. However, the difference was not
statistically significant.
    HS-II. The mean (+ SD) daily apparent absorption values were
153+77, 94+65 and 23+66 mg for 0.5, 1.7 and 2.9 g phytate intakes,
respectively. There was a significant difference between all
means $P < 0.05$, by ANOVA. A plot of apparent absorption of calcium
vs the dietary phytate/calcium molar ratio is shown in Figure 1.

Table III. Calcium Metabolic Balance Data, HS-I

Balance Component[1]	Dietary Treatment	
	Whole Bran Muffins	Dephytinize Bran Muffins
A. Days 1-5		
Intake	1070 + 94	1110 + 138
Fecal Exc.	912 + 191	818 + 165
App. Absorpt.[2]	161 + 138[4]	289 + 173*[4]
B. Days 6-15		
Intake	1100 + 121	1110 + 163
Fecal Exc.	729 + 117	742 + 159
App. Absorpt.	371 + 92	369 + 151
Urine Exc.	162 + 64	185 + 80
Balance[3]	208 + 58	184 + 87[4]

Note: Ten subjects.
[1]Values are means + SD, mg/day.
[2]Intake minus fecal excretion.
[3]Intake minus (fecal + urinary excretion).
[4]Negative value for 1 subject.
*Significantly different from whole bran muffins P< 0.05, Students paired t.

Because phytate intakes were essentially constant, but calcium intake varied with caloric needs, a range of phytate/calcium molar ratios resulted with each phytate level. The correlation coefficient for apparent absorption and dietary phytate/zinc molar ratio was 0.62, P < 0.01. The equation of the regression line was y = 181 + (-634)x, with a pronounced negative slope as the molar ratio of phytate/calcium increased. When the higher level of phytate was consumed, phytate/calcium molar ratios between 0.2 and 0.3, 6 of the 12 subjects excreted more calcium in the feces than was consumed. All subjects had positive apparent absorption values when consuming the low and intermediate phytate levels, phytate/calcium molar ratios of about 0.17 or less. Within the 15-day diet treatment, apparent absorption for the three consecutive 5-day periods was 94, 123, 64 when the intermediate level of phytate was consumed and 26, 39, 3 when the higher level was consumed. The ANOVA indicated no significant differences in these responses.

Apparent absorption of calcium for the six individuals in HS-III is shown in Figure 2. Only one of the six individuals, indicated by stars in Figure 2. tended to be fairly constant in

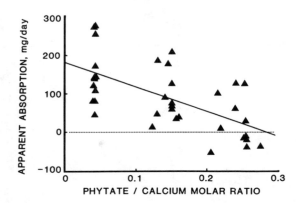

Figure 1. Dietary phytate/calcium molar ratio and apparent
absorption of calcium. Each triangle is the mean for one
individual for 3 consecutive 5-day menu cyles. See text for study
details, HS-II.

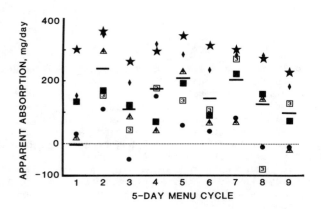

Figure 2. Apparent absorption of calcium for 9 consecutive 5-day
menu cycles with phytate/calcium molar ratio <0.05, HS-III. Each
individual is represented by a different symbol.

apparent absorption over the 9 periods. Some individuals varied considerably from one 5-day period to another, particularly the two individuals indicated by numbers. There was no consistent cyclical pattern across subjects.

Bulk liquids from the self-chosen diets of the volunteers in HS-II and HS-III have not been analyzed, however, milk was included with the 2nd plate food composites. Table IV is a summary of the distribution of calcium in the self-chosen food composites. Final

Table IV. Distribution of Calcium Content of Self-Chosen Diet Composites

Calcium Content	Percentage of Daily Composites
mg/day	
<500	14
500-1000	42
1000-1500	30
>1500	14

Note: 18 liquids, 231 total daily composites. Does not include bulk liquids such as water, soda, and beer.

of total intakes cannot be made until the bulk liquids are analyzed, however, almost one-half of these individual daily food composites contained 125% or more of the adult RDA for calcium. The phytic acid in the daily food composites ranged from 0.35 to 2.73 g/day (9). This information about self-chosen diets will provide a framework within which the practicality of the intakes and subsequent conclusions from the two controlled human studies may be assessed.

Discussion

For HS-I, with calcium intakes of about 1100 mg/day, no difference was observed in either apparent absorption or balance of calcium over the last 10 of 15 days when the phytate intake was 0.2 or 2.0 g/day. The molar ratio of phytate/calcium was either 0.01 or 0.1 in HS-I. In HS-II the calcium intake was lower, about 740 mg/day, but the same across three levels of phytic acid, 0.5, 1.7 and 2.9 g/day. The phytate/calcium molar ratios were 0.04, 0.14 and 0.24. Apparent absorption of calcium for the 15-day diet treatment period became progressively less as the molar ratio of phytate/calcium increased, to the extent that 6 of 12 individuals excreted more calcium in the feces than they consumed when the mean ratio was 0.24. About 200 mg of calcium was excreted daily in the urine by

subjects in HS-I. Urinary excretion for HS-II has not been determined, but daily urinary excretion of 100 mg will place most of the
individuals in negative calcium balance when phytate intake was
either the intermediate or high level. The phytate/calcium molar
ratio of 0.14 from HS-II is not much different from 0.11, the high
ratio in HS-I, yet the individuals seemed much more at risk of
negative calcium balance in HS-II. The difference may be the
additional 300 mg of calcium consumed in HS-I. Another possibility
is that the habitual calcium intakes of our volunteers may be
nearer 1100 mg and a number of the subjects may exhibit negative
balances when consuming 700 mg regardless of any additional
bioavailability influences from the diet.

The brown or whole meal bread diets employed by previous
investigators were often variable in calcium and phytate intakes,
not only between individuals, but by the same individual subjected
to different diet treatments. Nevertheless an estimate of the
molar ratio of phytate/calcium in the brown or whole meal bread
diets used by McCance and Widdowson (10), Walker et al. (11) and
Reinhold et al. (2, 12) is 0.25 or greater. These investigators
observed either negative or less positive calcium balance and
apparent absorption when the brown bread diets were consumed
compared to white bread diets with phytate/calcium molar ratios
less than 0.05. Our results support their findings. Reinhold et
al. (2) and McCance and Widdowson (13) used sodium phytate in some
studies as well as whole wheat bread and observed similar results.

Data from HS-II are incomplete, but unless urinary calcium
excretion was greatly decreased when the higher phytate levels were
consumed, a large fraction of our subjects will be in negative
calcium balance. McCance and Widdowson (13) observed a reduction
in urinary calcium for most subjects when sodium phytate was
incorporated into white bread and the phytate/calcium molar ratio
exceeded 0.3. Urinary calcium response to sodium phytate ingestion
was variable in the three subjects studied by Reinhold et al. (2).
Ingestion for 4 weeks of rice bran containing phytate decreased
urinary calcium in hypercalcuric patients (13). In another group
of hypercalcuric patients, some did not respond to treatment with
wheat bran and those that did respond reverted to higher urinary
calcium 1 month following cessation of bran treatment (14).

Walker et al. (11) interpreted changes in calcium balance of
their subjects with time, to indicate adaptation to ingestion of
the whole wheat bread. Reinhold et al. (15) observed no evidence
of adaptation by two men to unleavened bread consumed for 98 days.
The change between the first 5 and subsequent 10 days in apparent
absorption we observed in HS-I when the men consumed the whole bran
muffins might be interpreted to indicate adaptation. More likely,
there was simply a change in endogenous secretion of calcium into
the intestine to reestablish homeostatis if there was a drastic
decrease in absorption of calcium. There was possibly an indication of homeostatic adjustments with intermediate and high
phytate intakes in HS-II, but 15 days is probably too short a time
period to demonstrate adjustment. Phang et al. (16) studied the
effects of dietary perturbation of calcium metabolism in normal
man and reported that changes in parameters involving kidney and
intestine were in a direction to support calcium homeostasis.

HS-III (Figure 2) showed the wide variation in response between individuals to almost identical dietary conditions. Previous dietary history may be partially responsible for the differences. We have not completed the analyses of self-chosen diets consumed by the subjects in HS-II and HS-III, but preliminary indications are that there was a wide range in intakes of calcium and of phytate/calcium molar ratios. Stools and urine were collected while the self-chosen diets were being collected. Comparison of calcium metabolic balance parameters when consuming the self-chosen diets and the controlled diets of HS-II may provide an insight on dietary calcium requirement by these individuals. Some advocate the present RDA of 800 mg for adult men is too low (17), however, unpublished (R.D. Reynolds et al.) data from dietary intakes in a developing country indicate much lower usual intakes of calcium with relatively high phytate/calcium molar ratio.

Our studies do not resolve the question of phytate vs fiber for the effect of wheat bran on dietary calcium bioavailability. Phytate level clearly affected apparent absorption of calcium in HS-II in the presence of an amount of the water insoluble fraction of dephytinized bran equivalent to 12 g of untreated bran and the phytate supplied as sodium phytate. An additional trial using untreated bran and the same amount of fiber as the water insoluble fraction with sodium phytate could resolve the question of fiber vs phytate. In HS-I, the balances were positive when a relatively large amount of bran, 36 g/day, was consumed. Calcium intakes were possibly higher than most men consume, but under the dietary conditions imposed for 15 days, the phytate and fiber of 36 g of bran did not express an adverse effect on calcium balance. Possible future studies include chemical characterization of the fiber components of the dephytinized bran and the water insoluble fraction and in vitro binding studies using whole bran and the fractions (18).

Acknowledgments

We thank Priscilla Steele and Sheryl Cottrell of the Human Study Facility, BHNRC, Dr. Phylis Moser and Taryn Moy of the University of Maryland and David Hill of VMNL, BHNRC for their technical expertise in conducting these studies. We also thank Robert Maneval, Robert Crawford and Dr. Vernon Youngs, USDA, Spring and Durum Wheat Quality Research Laboratory, Fargo, ND, for procuring the wheat and milling the bran.

Literature Cited

1. Allen, L.H. Am. J. Clin. Nutr. 1982, 35, 783.
2. Reinhold, J.G., Nasr, K., Lahingarzadeh, A., Hedayati, H. Lancet 1979, 1, 283.
3. Morris, E.R., Ellis, R. In: Inglett, G.E. "Nutritional Bioavailability of Zinc" ACS Symposium Series 210, American Chemical Society, Washington, DC 1983, p. 159.
4. Morris, E.R., Ellis, R. J. Nutr. 1980, 110, 2000.
5. Ellis, R., Morris, E.R., Philpot, C. Anal. Biochem. 1977, 77, 536.
6. Ellis, R., Morris, E.R. Cereal Chem. 1983, 60, 121.

7. American Association of Cereal Chemists. "Approved Methods of the American Association of Cereal Chemists", AACC, St. Paul, MN, 1977, #32-30.
8. National Research Council "Recommended Dietary Allowances", National Academy of Sciences, Washington, DC, 1980.
9. Ellis, R., Morris, E.R., Hill, A.D., Kelsay, J.L. Fed. Proc. 1984, 43, 851 (Abstr.).
10. McCance, R.A., Widdowson, E.M. J. Physiol. 1942, 101, 304.
11. Walker, A.R.P., Fox, F.W., Irving, J.T. Biochem. J. 1948, 42, 452.
12. Reinhold, J.G., Abadi, P., Faraji, B., Ismail-Beigi, F., Russell, R.M. Baroda J. Nutr. 1980, 7, 55.
13. Ohkawa, T., Ebisano, S., Kitagawa, M., Morimato, S., Mizazaki, Y. J. Urolo. 1983, 129, 1009.
14. Shah, P.J.R., Green, N.A., Williams, G. Br. Med. J. 1980, 2, 426.
15. Reinhold, J.G., Faraji, B., Abadi, P., Ismaíl-Beigi, F. Ecol. Food Nutr. 1981, 10, 169.
16. Phang, J.M., Berman, M., Finerman, G.A., Neer, R.M., Rosenberg, L.E., Hahn, T.J., Fisher, L., Granger, A. J. Clin. Invest. 1969, 48, 67.
17. Spencer, H., Kramer, L., Lesniak, M., Bartolo, M., Norris, C., Osis, D. Clin. Orthopaed. Rel. Res. 1984, 184, 270.
18. Lyon, D.B., Am. J. Clin. Nutr. 1984, 39, 190.

RECEIVED December 26, 1984

Urinary Acid and Calcium Excretion
Effect of Soy Versus Meat in Human Diets

T. R. WATKINS, K. PANDYA, and O. MICKELSEN

School of Health Sciences, City University of New York, New York, NY 10010

Omnivores excrete a more acidic urine than herbivores.
Investigators have demonstrated increased urinary
acid and calcium by feeding acid-producing chemicals.
Such acid and calcium excretion may be important in
development of osteoporosis. To test diets of meat
and vegetable protein upon urinary acid and calcium,
nine human adults, aged 22 to 69 years, were fed
isonitrogenous diets of chicken or soy beans in
seven-day feeding periods. Diets provided daily:
80 g of protein; 450 mg calcium; 2300 mg sulfur; and
1900 mg phosphorus. Twenty-four hour urines were
collected in the prediet period, day 0, and days 2, 5,
and 7 of each diet period. Significant increases
occurred in the meat versus the soy period in titratable
acid, 38% (P < 0.02); calcium, 18% (P < 0.05); and,
sulfate, 6.0% (P < 0.05). Increased urinary calcium
after meat consumption could represent increased
buffering of alkaline bone salts, a defense against
the titratable acid load of the acid ash diet. The
implications of the data are discussed in terms of
the reduced bone mineral loss observed in vegetarians
compared with omnivores.

Bone is the main source of calcium in the human body.
Osteoporosis, decreased calcium salt reserves in the body, has become
the most prevalent bone disease in the U.S., being especially prevalent among post-menopausal women (1). Typical signs of this
debilitating condition include backache with spasms, wedge fractures
of the dorsal and lumbar vertebrae, and hip fractures (2).
Osteoporosis eventually is manifested as the "dowager's hump," with
concomitant loss of height in the seventh and later decades. From
15 to 50% of the women between 45 and 90 years living in nursing
homes have some osteoporosis. Populations noted to have a high
incidence of osteoporosis reside in western, generally industrialized,
societies.
    In the United States osteoporosis afflicts 26% of the Caucasian
female population, and they have vertebral crush fractures by age

0097–6156/85/0275–0073$06.00/0

65; of these women, 35 - 40% have at least one hip fracture by
age 90 years (3). A similarly high incidence has been documented
in the United Kingdom (4). Less developed countries have a markedly
lower incidence of osteoporosis. In his studies in South Africa,
Solomon (5) observed that rural Bantu women showed one-tenth as many
hip fractures as Caucasian women. Walker and his co-workers (6)
doing further studies of these groups noted that rural Bantu women
secured from their diet only half the calcium (250 - 400 mg/d) of
the Caucasians (800 - 900 mg/d). This was true even into the ninth
and tenth decades. The remarkable difference in bone health and
reduced fracture incidence existed among the Bantu women in spite
of the fact that they bore six children and breast fed them for as
long as two to three years; in contrast, the white women bore only
two children and fed them formula diets. Walker and his colleagues
initially were at a loss to explain the differential incidence of
osteoporosis among the white and Bantu women in South Africa. It
was suggested (7) that the vegetarian diet consumed by the Bantu
and the presence of meat in the diets of the white women probably
explained the skeletal loss of calcium from, and the development
of osteoporosis among, the white women.

In contrast to the Bantu, consumption of high-meat diets by the
North American Eskimos has been accompanied by severe osteoporosis.
Mazess and Mather (8) measured bone densities of both male and female
Eskimos of all ages. As early as the fourth decade of life, Eskimo
women had bones with less than 85% of the density of age, and sex-
matched white women living in the United States. Markedly larger
differences of bone occurred in later decades; this was true of
Eskimos of both sexes. The Eskimo diet, very high in protein, is
abundantly supplied with fish, reindeer, moose, caribou, and other
meats.

Recently, Marsh, et al. (9) have published a study of the bone
density of Seventh Day Adventists. In comparing the bone radius
densities by single photon absorptiometry of the Seventh Day
Adventists (S.D.A.) and age-, weight-, and sex-matched controls, they
found that in the sixth decade and later, during the post-menopausal
years, the average bone density of the women significantly exceeded
that of their controls. Their bone density stayed above the fracture
region, in contrast to the controls, whose bone density decreased
into the high risk fracture region by the sixth decade. The out-
standing characteristic of this S.D.A. group is its dietary avoid-
ance of meat products. Most adherents of that faith in southwestern
Michigan where these studies were done are lacto-ovo-vegetarians.

Age, calcium intake, hormonal status, exercise and vitamin status
have all been implicated in the development of osteoporosis. Estrogen
levels represent an important factor in skeletal calcium retention
and homeostasis. In therapeutic trials in which post-menopausal
women were given daily doses of estrogens, such therapy has been
demonstrated to be partially effective in reducing the rate of bone
resorption. However, this therapy has the concomitant hazard of
endometrial cancer (10). Vitamin D and its hormones have been given
considerable attention in the more recent studies. Without adequate
dietary and tissue levels of such vitamins, calcium absorption and
bone status will be impaired.

Exercise, too, has been reported to be an important factor in

maintaining bone calcium equilibrium. Bone density of manual
laborers markedly exceeded that of college professors and others
leading a more sedentary lifestyle (11).
     Numerous studies of dietary protein-induced urinary calcium loss
have appeared, in which the quantitative effect of protein upon
calcium has been investigated. Anand and Linkswiler (12) reported
that feeding a high-protein (150 g/d) diet to college students nearly
doubled their calcium output compared with their output when on a
low-protein (42 g/d) diet. Most of the calcium was lost in the urine
with only very small increases in fecal losses.

Animal Work

Many animal species excrete more calcium if fed an acid or acid-
forming compounds. In the calf, Steenbock and coworkers (13)
observed hypercalciuria and acidic urine after feeding hydrochloric
acid to the calf. Stehle (14) pointed out that calcium represented
the main long-term fixed base to be lost in the urine of the dog
loaded with excessive amounts of hydrochloric acid. Walzer and
Browder (15) demonstrated that when infused with a sulfate containing
solution, the dog excreted several fold more acid and calcium than
saline-infused controls; the increased calcium loss returned to
normal upon removal of the sulfate. Marone, et al. (16) demonstrated
increased excretion of calcium in the acidotic dog. Correction of
the acidosis reduced the excessive fractional calcium excretion rate,
but did not alter sodium excretion.
     The bone becomes depleted of calcium salts when the urine is
acidic over a relatively long period. This was shown by Goto (17)
who fed rabbits large doses of hydrochloric acid. He then showed
that urinary calcium loss occurred in concert with a marked reduction
in mass of the skeletal system, and also that the total non-fat
dry weight of bone decreased,implying a loss of bone matrix. A dose-
dependent, dietary acid induced loss of labelled calcium from rat
bone has been reported by Thorn and his coworkers (18). They
demonstrated that in response to graded doses of ascorbic acid, cells
in tissue culture, and bones in whole animals fed such doses were
depleted of the labelled calcium.
     The rat has been used rather widely to study the relation
between dietary protein, or acid salt feeding, and calcium loss.
Barzel and Jowsey (19) showed that the rat fed a control diet
supplemented with ammonium chloride excreted excessive urinary
calcium, and experienced a concomitant loss of fat-free bone tissue.
Draper, et al. (20) extending this work, reported an inverse relation
between dietary phosphate and loss of bone calcium and dry, fat-free
tissue. In subsequent studies (21), they reported that this process
was accompanied by reduced serum calcium levels; the high phosphorus,
low calcium diet increased urinary calcium loss. Whereas, increasing
the phosphorus content of the diet stopped the excessive urinary
calcium loss. To test possible zinc loss that might result from this
sort of acid salt feeding, Jacob and her coworkers (22) fed rats a
supplement of ammonium chloride and then measured urinary zinc and
calcium. The hypercalciuria occurred exclusive of an effect upon
urinary zinc loss.
     On the basis of these and other animal studies, the effect of

feeding diets having a highly acidic ash provided suggestive
evidence that humans consuming highly acidic diets according to the
observations of Sherman and Gettler (23) would also be expected to
excrete urine enriched in calcium. Rather than containing acid per
se, human diets usually contain the anion as a constituent of an
amino acid or as phosphate esters.

Human Diet Studies

Since acid-forming amino acids are present primarily in proteins,
a number of papers have reported the effect of feeding various levels
of protein to human subjects. Sherman and Gettler (23) postulated
that food ash determined the acidity or alkalinity of the urine.
Once a food was consumed and oxidized in the human body, the urinary
acidity would reflect the nature of the ash of the food as observed
in the rehydrated ash; care had to be taken to avoid losing sulfur
and chlorine during ashing. They pointed out that meats showed a
predominance of acid forming elements, milk tended toward milk
alkalinity, and the ash of fruit and vegetables was largely alkaline.
In a feeding study with one male subject, a calciuretic effect
prevailed when an acidic, rice-enriched diet was fed compared with
a potatoe diet. in studies with women, Bogert and Kilpatrick (24)
showed that feeding subjects an acidic diet resulted in elevated
urinary calcium, whereas an alkaline diet induced a decline in
urinary calcium.
    Acidosis induced by salt feeding to humans influenced urinary
calcium loss as effectively as feeding whole foods. Martin and Jones
(25), for example, fed adult subjects a diet supplemented with
ammonium chloride which resulted in marked hypercalciuria and an
acidified urine. In a follow-up trial, feeding alkali as sodium
bicarbonate, they also demonstrated that human hypercalciuria could
be prevented by adding an alkaline supplement to the diet.
    That increasing levels of protein in human diets result in
elevated urinary calcium loss has been reported by Margen, et al. (26).
These investigators pointed out that as the dietary protein level
increased, greater calcium supplementation would not assure positive
calcium balance, i.e., calcium retention. Extending these studies,
Allen and her coworkers (27) demonstrated the high protein effect on
calcium loss, even in longer-term feeding trials of 90 days in which
males were fed 225 g of protein per day in mixed formula diets,
compared with 75 g per day. The higher protein intake doubled the
calcium output in the urine.
    Oxidation of dietary sulfur results in a rather striking effect
upon the output of urinary acid and hypercalciuria. Zemel, et al.
(28) studied this in adult subjects. When whole food dietary sulfur
intake increased, so did urinary calcium. About half the effect of
the sulfur-containing protein was observed after feeding the subjects
an equivalent amount (on the basis of sulfur content) of the amino
acids cysteine and methionine. This emphasizes the importance of
testing whole food diets, as well as semi-synthetic ones in
investigating diet and hypercalciuria.
    In light of the higher bone densities and lower fracture rates
among persons eating little or no meat, and lower densities and higher
rates of fracture among those eating meat-rich diets, and the known

effect of high-protein diets upon calcium balance, we evaluated the
effect of two different proteins fed at the RDA level upon urinary
calcium loss. Both urinary acid and calcium excretion were studied
in healthy adults of both sexes fed a fixed diet with the protein
provided either by baked chicken or soy bean. Subjects ate equivalent
levels of protein, calcium, phosphorus, and sulfur during each of the
test periods of the study.

We were desirous of determining whether animal protein, so
abundant in the American diet, and a vegetable like soy bean, might
yield different results in terms of urinary acid and calcium loss,
even if fed at isonitrogenous levels. Soy beans were chosen for this
study since the sulfur content of the cooked soybeans was similar to
that of meat.

The Diet

The study was designed to compare the effects of feeding a mixed,
vegetable-based soy protein diet with a diet containing chicken.
The protein in both diets was at the level recommended by the
National Academy of Science. To maintain body weights, those
subjects requiring extra calories were given hard candies which
contained no protein. Twenty four hour urines were collected from
each subject on the day preceding the initiation of each study as
well as on the second, fifth, and seventh day of each dietary program.
The urine was preserved by toluene added to each collection bottle
and then refrigerated. Urinary titratable acidity and PH as well as
calcium balance during the two week feeding periods comprised the
main measures of response, along with sulfate excretion. The two
diets were controlled within three per cent for protein, sulfur,
phosphorus, calcium, magnesium, and calories. The basal diets
contained 2000 Calories, 80 g protein, 460 mg calcium, 1900 mg
phosphorus, 2300 mg sulfur, and 590 mg magnesium. Menus appear in
Table I, A & B. After the initial soy period, a four-day period
allowed for free-choice diet adjustment, the meat period followed.
During the diets no untoward effects were noted by the subjects.
Three subjects experienced some flatulence in the soy period; three
subjects lost 1 - 2 kg in that period.

The experimental diets were similar except for chicken and soy
beans. A batch of the latter was generously provided by Dr. M. Teel
of the Agronomy Department of the University. These beans were a
strain that when cooked, could be eaten in fairly large amounts
without any discomfort. To minimize dietary variation during the
study, similar foods were served each day except for the chicken
and soy beans. The chicken was secured from the University
commissary in a large lot which was sufficient to last throughout
the study. All foods were acceptable to the subjects who had no
complaints about their taste.

Methods

In the morning, after the subjects delivered the urine samples, pH,
total titratable acidity (TTA), and their calcium and sulfate contents

Table I.   A.   Composition of Soy Bean Based Vegetarian Diet

Food Item	Weight, g
Breakfast	
Orange juice, from concentrate	250
Apple juice, from concentrate	125
Wheat flakes, Wheaties	30
Wheat germ	14
Banana	50
Bread	50
Lunch	
Spaghetti, Ideal, cooked	200
Spaghetti sauce, Ragu	115
Bread	50
Margarine	5
Romaine lettuce	50
Apple	50
Dinner	
Soy bean chili	300
Bread	100
Rice, brown, cooked	50
Margarine	5
Romaine lettuce	100
Tomato, fresh	50
Cucumber, unpared	100
Wheat germ	14
Banana	100
Orange, weight with skin	100

B.   Composition of Chicken Based Meat Diet

Food Item	Weight, g
Breakfast	
Orange juice, from concentrate	250
Oatmeal cereal, Quaker's	35
Milk, lowfat	60
Egg, large, hard boiled	50
Roll, sweet	80
Bread	50
Margarine	5
Lunch	
Soup, lentil, Progresso	100
Salad, macaroni	50
Bread	50
Jello with mixed fruit	150
Apple	100
Dinner	
Chicken, boneless, baked	100
Potato, baked	200
Rice, brown, cooked	50
Cabbage, cooked	100
Romaine lettuce	100
Cucumber, unpared	100
Tomato, fresh	50
Salad dressing, Italian, regular	15
Margarine	5
Banana	100
Orange, weight with skin	100

were determined. TTA was measured by titration of a 25 ml urine
aliquot with standardized 0.1N sodium hydroxide to a bromothymol blue
endpoint. Calcium was measured by atomic absorption spectrometry
after appropriate dilution of the urine in 0.5% lanthanum oxide (29).
Sulfate was measured spectrophotometrically as barium sulfate
according to the method of Ma and Chan (30). Statistical analysis of
paired differences were done for data of both test periods (31).

The Subjects

Four women and five men, aged 22 to 69 years, from the student body
and faculty of the University, volunteered to participate in the
study as approved by the Human Subjects Committee. One woman had
passed her menopause; her husband was retired; the other subjects
were graduate students or young faculty members. Most of the subjects
led relatively sedentary lives except for one student who was engaged
in a physical activity program and another who spent part of each
weekend riding his bicycle in preparation for a bicycle race. The
subjects, all in good health, except for one with mild hypertension,
ate their meals in the metabolic unit on the campus. Weights of the
subjects ranged from 47 to 78 kg.; heights from 158 to 180 cm. None
of the subjects were taking any medication routinely.

Results

According to food composition tables, the diets used in the study
provided an adequate intake of the essential nutrients. Each day,
an exact amount of each food served the subjects was saved for analysis
of nitrogen, calcium, and sulfur (as sulfate). The results for these
elements agreed with those in the literature. These results indicated
that during the soy diet, mean intakes were: 78.5 g protein, 456 mg
calcium, and 2305 mg. of sulfur. Intakes in the meat period were
within 3% of those of the soy period.

Urinary pH and Titratable Acidity

Urinary pH varied only slightly for the two dietary periods. During
the soy feeding period, the pH values ranged from 5.8 to 6.9; in
the meat period, from 5.6 to 6.9.

     The total titratable acidity (TTA) of the urine varied
significantly between the two dietary periods, $P < 0.02$. During the
soy period, the TTA ranged from 10 to 50 mEq/d. Meat consumption
was accompanied by an increase in the TTA values which ranged from
15 to 70 mEq/d, Figure 1.
     Compared with the TTA during the prediet period, the soy bean
diet resulted in an increase by day 2, with the same value for TTA on
day 5. Thereafter, there was a slight decrease on day 7. The TTA
values for day 2 ranged from 12.7 to 44.6 mEq/d. Why there should
be this 3.5-fold spread in the excretion of titratable acid for
subjects consuming the same diet is unknown.

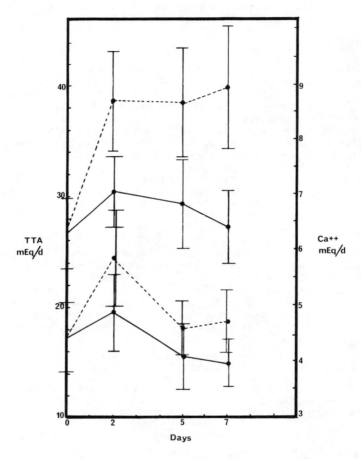

Figure 1. Effect of soy bean and meat diets on average urinary total titratable acidity (TTA) and calcium. Upper two curves represent TTA; lower two curves, urinary calcium. Bars indicate S. E. M. Key: broken line, meat; and solid line, soy.

Although the average TTA value for day 5 of the soybean period was
slightly lower than that for day 2, 3 subjects experienced an
increased urinary acid excretion on day 5.

When baked chicken replaced soy bean, the TTA values increased
about 40% above the baseline value. In only three of the nine
subjects did the TTA values decrease by the fifth day. In six of the
nine, TTA values continued to increase even on the seventh day of
the diet. Values ranged from 15.0 to 53.3 mEq/d on day two in
contrast to the soy data. Although no significant sex-specific, or
age-specific response could be detected, the maximum response of
older subjects occurred on day five, whereas younger ones had peaked
somewhat earlier, on day two.

Statistical comparison of the mean differences confirmed
significant changes from the prediet to the soy values, and to the
meat values. A 4% increase in measured mean TTA occurred in soy
versus prediet, whereas a 46% increase in TTA was measured in the
meat versus prediet values.

Urinary Calcium Excretion

When subjects ate the meat diet, mean calcium losses in the urine
increased significantly in all subjects (P<0.02). In the soy period,
the daily urinary calcium loss averaged from 1 to 7.5 mEq with an
average of 4.23, but in the meat period, the daily loss ranged
from 2 to 10.5 mEq with an average of 5.07.

In six subjects, urinary calcium excretion during the soy period
peaked by day 5, and began to decrease, but in three subjects, it
began to decline by day 2. Two subjects did not show any rise in
urinary calcium. In one subject the value on day 5 was lower than
that on day 2; thereafter, it resumed its upward trend. For all
subjects, the kinetics of the rise and fall to near initial values
suggested that an equilibrium may have been established.

During the meat diet period, calcium excretion of six of the
nine subjects reached a maximum by day 2. Yet in three subjects
calcium excretion continued to increase till day 7. From these
data, one may infer that an equilibrium probably had not been
established during this seven-day period.

Mean values of urinary calcium during meat eating exceeded the
soy values at comparable time points for all subjects except one.
This subject had a greater calcium loss during the soy than the
meat feeding period.

Statistical analyses of the calcium excretion data of the prediet, soy, and meat periods revealed significant differences. The
paired differences between the calcium loss in urine during meat
and soy diets was significant at the P<0.05 level.

Urinary Sulfate Excretion
Consumption of the meat at isonitrogenous and isosulfurous levels
compared with the soy beans resulted in modestly increased sulfate
excretion Figure 2. When soy was fed, the subjects excreted
sulfate that ranged from the prediet value of 49.3 to 58.9 mEq/d by
day 7. With meat feeding, each point sampled was slightly greater
than in the soy period, the day 7 value having risen to 61.9 mEq/d.
That the curves of the sulfate loss for each protein source had

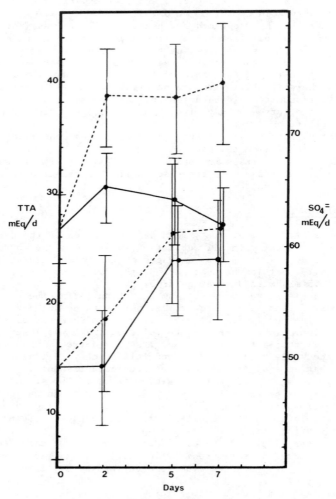

Figure 2. Effect of soy bean and meat diets on average urinary total titratable acidity (TTA) and sulfate. Upper two curves represent TTA; lower two curves, sulfate. Bars indicate S. E. M. Key: broken line, meat; and solid line, soy.

parallel shapes suggested that no large differences in rate of digestion, absorption, or excretion occurred as a result of the change in protein source. The enhanced excretion of sulfate that followed meat eating represented statistically significant paired differences (P<0.02).

The similarity in sulfate excretion during the meat and soybean periods (Figure 2) indicates that the two diets were well matched for sulfur content. Despite this similarity, the titratable acidity of the urine during the soy bean period was far below that of the meat period and did not follow the sulfate excretion. This suggests that some other anion other than sulfate in the titratable acid may have decreased while sulfate rose, Figures 1 and 2.

Age

No significant differences were detected in the excretory patterns of TTA, calcium, or sulfate based upon age. Yet the TTA was generally more than 20% higher among older subjects for either diet than for younger ones. During the soy period no significant change could be measured for older versus younger participants. Also older subjects excreted about 12% less calcium in the urine than the younger ones. This occurred despite the fact that the calcium intake of all subjects was the same.

Sex

No marked differences in dietary influence upon acid or calcium excretion or sulfate loss was observed in males compared with the females in this study.

Discussion

The meat diet resulted in markedly greater titratable acid and calcium excretion compared with the soy diet (P<0.02). This occurred despite the fact that each diet contained the same amounts of protein, calcium, phosphorus, and sulfur. Increased urinary calcium excretion in subjects accompanied this increased output of TTA (P<0.02).

Renal excretion of calcium is related to acid excretion. The evidence therefore is: First, the prediet and test diet urinary acidity and calcium excretion followed a similar time course. Each reached a peak on day 2 for the soy feeding period, and day five for the meat diet. Further, individual data points for each period (data not shown) also showed such similarities in magnitude and direction. Second, withdrawal of the soy protein resulted in a reduction of TTA and calcium to prediet values. Third, the magnitude of the change in acid and urinary calcium output though different, occurred in a similar range. Fourth, since the time course of acid excretion occurred more slowly in the meat period, not reaching a peak until day 5, and the calcium loss followed within hours, the evidence suggests more than a coincidental relationship.

Earlier animal work showed similar results in terms of urinary acid production from dietary precursors that could be converted into acid before excretion. However, most investigators used salts rather than foods containing the anion or its precursor. The addition of acid, in the form of hydrochloric, sulfuric, or ammonium chloride, acid phosphate salts, or ascorbate resulted in enhanced urinary acidity and concomitant calcium excretion. For example, in the detailed study of bone salt metabolism, Barzel and Jowsey (19) showed that the rat fed supplementary ammonium chloride subsequently lost more calcium, and developed markedly demineralized fat-free bone mass.

Our studies show that sulfur feeding in the form of meat also induced greater acid and calcium excretion in the urine than soy. Meat consumption was accompanied by more sulfate output in the urine. This agrees with earlier feeding studies in which animals were fed any of several fixed acids which had been reported to result in increased calcium loss (13, 17). Feeding hydrochloric acid, Goto (17) showed that calcium carbonate was simultaneously lost from the bones of rabbits. A similar conclusion was drawn by Steenbock, et al. (13) who demonstrated a loss of both calcium and phosphorus mediated by hydrochloric acid. These data were consistent with the sulfate infused dog. In the dog, Walzer and Browder (15) showed that sulfate feeding increased calcium losses along with sulfate loss. This confirmed results that Lamb and Evvard (32) reported for the pig.

Acidosis resulting from endogenous acid production also induces urinary calcium loss. The kidney responds to such acid production, as in the case of dietary acid, by excreting more calcium. Marone, et al. (16) demonstrated this point with the acidotic dog. When the acidosis was induced by ligating one kidney, leaving a remnant kidney, the ensuing acidosis was always accompanied by increased excretion of calcium.

The elevated urinary acidity in the chicken period may have resulted from, among other factors, the distinct sulfur amino acid ratios. The ratio of Met/Cys (mg/g N) in chicken is 150/80; whereas in soybeans it is 80/100, or in more comparable terms 150/100 suggesting that in relation to methionine, soybeans have twice as much cysteine as chicken. Sherman and Gettler (23) noted the importance of both phosphorus and sulfur. That sulfur containing compounds in the diet have an important effect on urinary acidity was confirmed by Lemann et al. (39) who demonstrated that elevated urinary acidity accompanied methionine feeding in humans.

The high biological value of soy has been attributed to its sulfur amino acid content. As mentioned previously, the sulfur amino acid ratios may have accounted for the large change in acid excretion in the meat period compared with the soy (Figure 1). Since this increased calcium loss, the quality of dietary protein may be important in conserving body calcium in the bone reservoir via the kidney. Human renal studies have corroborated animal data in-so-far as calcium excretion as influenced by urinary acidity is concerned. This was emphasized by Marone et al. (15) who reported increased excretion of calcium in the acidotic dog and by Zemel, et al. (27) who studied calcium filtration by the kidney. They fed subjects low or high-protein (50 or 150 g/d) diets, then compared

this with the effects of the low protein diet plus sulfur amino
acids in such amounts to make the low protein diet comparable to
the high protein diet in sulfur content. Thereby they accounted
for about half of the calcium loss produced by the high protein diet.
Their explanation was that the fractional reabsorption of calcium
by the kidneys decreased in the high protein period.

   None of these earlier animal studies had been rigorously for
protein, calcium, or acid output. The data of the present study
demonstrated that the acid nature of the diet, fed at constant levels,
and to the same subjects, induced hypercalciuria in the meat period.

   This phenomenon also had been reported in human subjects fed
acid ash foods. Farquharson, et al. (33) fed a high-protein (200 g)
diet to human subjects who promptly excreted more urinary acid and
calcium. This occurred whether the protein level was raised to
200 g, or an equivalent amount of ammonium chloride was fed. If,
on the other hand, the acid ash in the protein were neutralized with
sodium bicarbonate, the hypercalciuria did not occur.

   Feeding human subjects meat-rich diets has been clearly and
repeatedly associated with elevated urinary acid and calcium loss.
Anand, et al. (12) fed college men low, medium, and high-protein
diets. Fecal calcium changed less than 15% , whereas urinary
calcium losses accounted for the major loss of calcium. Calcium loss
and meat protein intake correlated directly. Johnson, et al. (34)
reported that women showed a similar response when fed such diets.

   Dietary phosphorus exerts variable influence on calcium loss
depending on the nature of the dietary protein. Humans fed food
containing abundant phosphorus to calcium excreted little more
calcium unless the meat content of the diet changed markedly.
Spencer, et al. (35) measured no large increased urinary or fecal
calcium loss in subjects fed a high phosphorus (meat) diet implying
a sort of protection imparted by the phosphorus. Similarly, in work
of Schuetté and Linkswiler (36) high phosphorus relative to calcium
intake actually resulted in marginal retention of calcium since no
difference in phosphorus intake occurred in either the meat or soy
diets of the present study. The larger calcium loss during the meat
period was attributed to some other property of meat versus soy
protein than the phosphorus to calcium ratio, per se. The effect
of the meat was principally in increasing acidity of the urine
which in turn resulted in extra loss of calcium.

   In yet another study, Hegsted, et al. (37) found that feeding
a high protein level occasioned an increase in urinary calcium.
Eating such a diet elevated titratable acidity and calcium loss in
the urine. A similar relation was observed in postmenopausal women.
Lutz and Linkswiler (38)fed eight subjects low (50 g/d) or high
(110 g/d) protein diets in a crossover pattern for 15 days. During
the high protein period, calcium excretion rose from 90 to 172 mg/d;
calcium absorption improved by 10%. No significant change in bone
density was observed.

   These data about the effect of dietary protein and calcium loss
are consistent with epidemiological evidence linking diet and bone
calcium loss. Walker, et al. (6) reported fracture rates ten times
greater among white South Africans eating a meat-rich European type
diet than among the local Bantu who consumed a vegetarian diet.
This pertained despite calcium intakes of 200-300 mg/d among the

Bantu women and 800-900 mg/d among the white women. In the United States, Marsh, et al. (7) reported significant differences in bone mass in age and height-matched lacto-ovo vegetarian Seventh Day Adventist women compared with omnivores. This was true in the women over 60 years of age; premenopausal S.D.A. women had bone densities which were similar to those of white omnivorous women. The lacto-ovo vegetarian diet that these S.D.A. women follow would produce a slightly alkaline urine compared with the acidic urine associated with the consumption of a meat diet.

There are a number of factors influencing bone density, especially in women who have passed the menopause. One of these is the nature of the diet apart from its calcium content. Little attention has been given to this means of maintaining bone mineralization especially in relation to osteoporosis. Yet the data from the S.D.A. women suggests the lacto-ovo-vegetarian diet is an important factor in maintaining bone mineralization in women who have passed the menopause. The more densely mineralized bones of these elderly S.D.A. women occurs despite the fact that their bone mineralization prior to the menopause is the same as that of omnivorous white women as shown by cross-sectional studies (9). This suggests that it may be advisable for women approaching the menopause to decrease their intake of meat, fish, and poultry. At what age that should be done and how drastically the reduction should be has not been established. The evidence from the present study and the reports in the literature suggest that a change in the nature of the diet is followed almost immediately by an alteration in the calcium lost from the body. On that basis, it may be possible to postpone a dietary change until the first signs of menopause appear.

Acknowledgments

We are indebted to Dr. Merle Teel for graciously providing the soy beans.

Literature Cited

1. Albanese, A. Bone Loss: Causes, Detection, and Therapy. A. R. Liss, New York, 1977.
2. Gruber, H. E.; Vaylink, D. J. J. Am. Geriat. Soc. 1981, 29, 490-497.
3. Gallagher, J. C.; Melton, L. J.; Riggs, L. B.; Bergstrath, E. Clin. Orthoped. 1980, 150, 163-171.
4. Wachman, A.; Bernstein, D. S. Lancet 1968, i, 958-959.
5. Solomon, L. J. Bone Joint Surg. 1968, 50B, 1.
6. Walker, A. R. P.; Walker, B. F.; Richardson, B. D. Am J. Clin. Nutr. 1974, 23, 244-245.
7. Mickelsen, O. In "Nutrition, Longevity, and Aging"; Rockstein, M. and Sussman, M., Eds.; Academic: New York, 1976, 123-142.
8. Mazess, R. B.; Mather, W. Am. J. Clin. Nutr. 1974, 27, 916-925.

9. Marsh, A.; Sanchez, T. V.; Mickelsen, O.; Kerser, J.; Mayor, G. J. Am. Diet. Assoc. 1980, 76, 148-151.

10. Monologas, S. C.; Anderson, D. C.; Lindsey, R. Lancet 1979, ii, 597-600.

11. Doyle, F.; Brown, J.; Lachance, C. Lancet 1970, i, 391-393.

12. Anand, C. R.; Linkswiler, H. M. J. Nutr. 1974, 104, 695-700.

13. Steenbock, H.; Nelsen, V. E.; Hart, E. B. J. Biol. Chem. 1915, 19, 399-419.

14. Stehle, R. L.; McCarty, A. C. J. Biol. Chem. 1921, 47, 315-319.

15. Walser, M.; Bowder, A. A. J. Clin. Invest. 1959, 38, 1404-1411.

16. Marone, C. C.; Wong, N. L. M.; Sulton, R. A. L.; Dirks, J. H. Nephron 1981, 28, 584-589.

17. Goto, K. J. Biol. Chem. 1919, 36, 355-376.

18. Thorn, P. J. Nutr. 1970, 100, 1479-1486.

19. Barzel, U. S.; Jowsey, J. "Osteoporosis"; I. Grune & Stratton: New York, 1978.

20. Draper, H. H.; Sie, T. L.; Bergan, J. G. J. Nutr. 1972, 102, 1133-1142.

21. Bell, H. R.; Engelmann, D. T.; Sie, T. L.; Draper, H. H. J. Nutr. 1975, 105, 475-483.

22. Jacob, M.; Smith, J. C.; Chan, J. C. M. Ann. Nutr. Metab. 1983, 27, 380-385.

23. Sherman, H. C.; Gettler, A. O. J. Biol. Chem. 1912, 11, 323-338.

24. Bogert, L. J.; Kirkpatrick, E. E. J. Biol. Chem. 1922, 54, 375-399.

25. Martin, H. E.; Jones, R. Am. Heart J. 1961, 62, 206-210.

26. Margen, S.; Chu, J.; Kaufmann, N. A.; Calloway, D. H. Am. J. Clin. Nutr. 1974, 27, 584-589.

27. Allen, L. A.; Bartlett, R. S.; Black, G. D. J. Nutr 1979, 109, 1345-1350.

28. Zemel, M. B.; Schuette, S.; Hegsted, M.; Linkswiler, H. J. Nutr. 1981, 111, 545-552.

29. Willis, J. B. Anal. Chem. 1961, 33, 556-559.

30. Ma, R. S. N.; Chan, J. C. M. Clin. Biochem. 1973, 6, 82-85.

31. Armitage, P. "Statistical Methods in Medical Research"; J. Wiley & Sons: New York, 1974.

32. Lamb, A. R.; Evvard, J. M. J. Biol. Chem. 1919, 37, 317-327.

33. Farquharson, R. F.; Salter, W. T.; Tibbets, D. M.; Aab, J. C. J. Clin. Invest. 1931, 10, 221-250.

34. Johnson, N. E.; Alcantara, E. N.; Linkswiler, H. J. Nutr. 1970, 100, 425-431.

35. Spencer, H.; Kramer, L.; Osis, D.; Norris, C. Am. J. Clin. Nutr. 1978, 31, 2167-2180.

36. Schuette, S.; Linkswiler, H. J. Nutr. 1982, 112, 338-349.

37. Hegsted, M.; Schuette, S.; Zemel, M.; Linkswiler, H. J. Nutr. 1981, 111, 553-562.

38. Lutz, J.; Linkswiler, H. Am. J. Clin. Nutr. 1981, 34, 2178-2186.

39. Lemann, J., Jr.; Relman, A. S.; Conners, H. P. J. Clin. Invest. 1959, 38, 2215-2223.

RECEIVED December 26, 1984

Mouse Bone Composition and Breaking Strength
Effect of Varying Calcium and Phosphorus Content in Animal or Plant Protein-Based Diets

NWEZE NNAKWE[1] and CONSTANCE KIES

Department of Human Nutrition and Food Service Management, University of Nebraska, Lincoln, NE 68583

The effect of varying calcium and phosphorus contents of animal or plant protein based diets on bone breaking strength and on calcium and phosphorus utilization of weanling mice was investigated. Sixty-four weanling mice were divided into eight groups of eight mice each and were fed rations containing two levels of calcium and phosphorus (0.3% and 1.2%) in the four possible combinations in egg white or soy based diets for eight weeks. Mice were individually housed in metabolism cages. Femurs of mice fed the animal protein diets exhibited more resistance to bone breakage than did animals fed the plant protein diets. Feeding of egg white protein in comparison to soy protein diets inhibited intestinal absorption of phosphorus as indicated by increased fecal phosphorus losses. Increased blood serum calcium levels were also observed in these mice. Feeding of low phosphorus diets also enhanced femur bone breakage resistance in comparison to feeding high calcium diets. The results suggested that in populations maintained largely on vegetarian diets the maintenance of a low phosphorus intake may be of greater importance than is the increasing of dietary calcium.

Osteoporosis, decrease in bone mass, and osteomalacia, decrease in bone mineralization, may occur in aging humans, sometimes concurrently within the same individual. Various nutritional, hormonal, iatrogenic, physical and circulatory factors have been implicated as influencing bone health (1).

Previous studies have shown that mice and rats are subject to osteoporotic changes in the skeleton, including cortical thinning and loss of mineral and organic matrix, which are analogous to

---

[1] Current address: Department of Home Economics, Illinois State University, Normal, IL 61761

those observed in humans. These changes occur in spite of the
presence of adequate amounts of calcium, phosphorus, vitamin D and
fluoride (2). However, the rate and extent of bone resorption are
enhanced by feeding diets which are either low in calcium or high in
phosphorus.

   Although much attention has been directed toward the relationship between calcium intake and osteoporosis, little consideration
has been given to the possible influence of dietary phosphorus on
the development of this disease in either man or animals. In a
study designed to determine the optimal concentration of calcium
and phosphorus in the diet of adult mice, aging animals were found
to undergo a greater loss of bone when the Ca/P ratio was 1:1 than
when it was 2:1 (3). The detrimental effect of the 1:1 ratio was
greatest when intakes of both calcium and phosphorus were high
(1.2% of the diet). These results implied that high phosphorus
diets exacerbate the loss of bone in aging mice by a mechanism which
cannot be counteracted by high intake of calcium.

   One explanation for the adverse effects of excessive phosphorus
intakes on bone health of rodents is that of secondary hyperthyroid-
ism (4,5,6). Under these circumstances, PTH activities and cAMP
urinary excretions would be expected to increase and blood serum
phosphorus would be expected to increase. This in turn would be
expected to result in formation of calcium-phosphorus complexes,
decrease in serum ionized calcium, parathyroid stimulation and bone
resorption.

   High intakes of dietary protein and/or phosphorus have been
implicated as possible factors related to the onset of osteoporosis
(8,9). Linkswiler and co-workers (10) and Johnson et al. (8) have
implicated high dietary protein as a possible contributory factor
in the etiology of osteoporosis because of increased urinary calcium
excretion and negative calcium balance of human subjects fed high
protein diets. However, the high urinary calcium losses caused by
feeding high protein diets were decreased when phosphorus intakes
were increased. Meat or purified sources of protein and calcium
were used in these studies. A decrease in urinary calcium excretion
should not be interpreted as being necessarily beneficial since it
may be indicative of mineralization of soft tissue, particularly
the kidneys, as a result of increased PTH activity (4,5,11,12,13).

   Meat and such high protein plant foods as soy are excellent
sources of phosphorus as well as protein. The phosphorus in meat is
readily absorbed from the gastrointestinal tract; however, much of
the phosphorus in plant products is in a bound form which may inhibit the absorption of calcium as well as phosphorus. This study was
designed to determine the effect of different levels of calcium and
phosphorus with plant protein or animal protein on bone breaking
strength and calcium and phosphorus utilization of weanling mice.

Experimental Procedures

Sixty-four C3H/HeN Sch inbred, female, weanling mice, weighing 10
to 13 grams, were purchased from Harlan-Sprague Dawley Co., Madison,
Wisconsin. Mice were fed mouse chow (Purina) for two days after
arrival to permit adjustment to the new environment and to allow
time for recovery from travel stress. The mice were then divided
into two groups. Group A mice were fed two levels of calcium
(0.3% and 1.2%) and phosphorus (0.3% and 1.2%) in the four possible
combinations with soy protein as the protein source and Group B mice

were given various levels of calcium and phosphorus with egg white protein as the protein source. Composition of the experimental diets is given in Table I. Rations were mixed according to specifications from TEKLAD Test Diets. The composition of the dietary vitamin and mineral mix are presented in Tables II and III.

The animals were housed in individual, wire-bottomed, stainless steel, metabolic cages. The animals were weighed and randomly assigned to eight treatment groups of eight mice each. Individual cages within a treatment group were randomly arranged on three cage racks in a room specifically used for animal experimentation. The room was maintained at 70 degrees Fahrenheit with 12 hours of artificial light alternating with 12 hours of darkness.

Mice were weighed on a weekly basis. Urine was collected in acid-washed containers for each seven days. Feces were collected daily and composited weekly. The animals were fed daily and feed intake was determined weekly by differential weighing. Glass-distilled water was given ad libitum.

Animals were maintained on the respective rations for a period of eight weeks. At the end of the eight weeks. the animals were randomly weighed, anesthetized with ethyl ether, and blood samples were collected. The bones (femur) were surgically removed. Analyses for calcium and phosphorus content of bone, feces, blood serum and urine were made using a Varian Techtron Model 1200 Atomic Absorption Spectrophotometer. Phosphorus content of these materials was measured by the method of Fiske and Subbarow (14) using a Beckman Spectrophotometer Model 24. Bone breaking strength was determined on fresh bone using an Instron Machine.

Analysis of variance was the statistical model used with pre-planned comparison testing for significant differences by the least square means method.

Results and Discussion

Femur bone breaking strength was used as a functional measurement of calcium-phosphorus nutritional status. Significantly greater ($P < 0.05$) bone breaking strength was exhibited by animals fed egg white protein than by animals fed soy experimental rations. Summary of effects of alterations in dietary protein, calcium and phosphorus sources in diets of young mice is presented in Table IV. Mean bone breaking strength of all mice receiving egg white protein was 8.19N and those receiving soy protein was 7.42N. Mean bone breaking strength of all mice receiving high phosphorus rations was 7.40N, and for those receiving low phosphorus rations was 8.21N. These data indicate that both low phosphorus and egg white diets contributed to the development of bone more resistant to breakage in comparison to soy protein diets and high phosphate diets as illustrated in Figure 1. There were no statistical effects of dietary calcium levels and source of protein. It should be pointed out that bone breaking strength was measured on fresh, not dried femurs; hence an increase in bone breaking strength might be the result of increased bone flexibility and bendability as would be the case in osteomalacia.

Mean femur bone weight of mice fed the soy base rations was higher than for mice fed the egg white base rations (0.039 g versus

Table I. Composition of Rations

Component %	Soy Based Rations				Egg White Based Rations			
	0.3% Ca 0.3% P	1.2% Ca 1.2% P	0.3% Ca 1.2% P	1.2% Ca 0.3% P	0.3% Ca 0.3% P	1.2% Ca 1.2% P	0.3% Ca 1.2% P	1.2% Ca 0.3% P
Soy assay protein	20.21	20.21	20.21	20.21	--	--	--	--
Egg white solids	--	--	--	--	21.88	21.88	21.88	21.88
DL-methionine	0.25	0.25	0.25	0.25	--	--	--	--
Biotin	--	--	--	--	0.0004	0.0004	0.0004	0.0004
Sucrose	53.1816	48.725	49.6887	50.9316	51.3845	46.976	48.6207	49.1343
Corn starch	15.00	15.00	15.00	15.00	15.00	15.00	15.00	15.00
Corn oil	5.00	5.00	5.00	5.00	5.00	5.00	5.00	5.00
Fiber (cellulose)	3.00	3.00	3.00	3.00	3.00	3.00	3.00	3.00
Vitamin mix[1]	1.00	1.00	1.00	1.00	1.00	1.00	1.00	1.00
Mineral mix[2] (Ca and P deficient)	1.337	1.337	1.337	1.337	1.337	1.337	--[3]	1.337
Calcium phosphate, monobasic	0.6214	4.2837	1.6828	0.6214	1.133	4.796	1.80	1.133
Calcium carbonate	0.40	1.1943	--	2.65	0.2658	1.059	--	2.515
Potassium phosphate, monobasic	--	--	1.5497	--	--	--	1.9163	--
Sodium phosphate, monobasic	--	--	1.3366	--	--	--	1.3366	--

Note: Rations were mixed according to specifications.

[1] Composition of the vitamin mix from TEKLAD (#40060) is given in Table II.

[2] Composition of calcium and phosphorus deficient mineral mix from TEKLAD (#170913) is given in Table III.

[3] In order to balance mineral intake, the following minerals were used in place of the usual mineral mix (% of diet): potassium chloride, 0.32; magnesium oxide, 0.084; manganese carbonate, 0.0123; ferric nitrate, 0.021; zinc carbonate, 0.0056; cupric carbonate, 0.0011; potassium iodate, 0.0011; potassium iodate, 0.0004; sodium selenite, 0.00003; chromium potassium sulfate, 0.00193.

Table II. Vitamin Mix (TEKLAD) (Catalog #40060)

Contents	g/Kg of vitamin mix	Supplies in grams or units per Kg of diet when added at 12% or 10g/Kg of diet
P aminobenzoic acid	11.0132	0.110132
Ascorbic acid, coated (97.5%)	101.6604	1.016604 = 0.9912 g ascorbic acid
Biotin	0.0441	0.000441
Vitamin B12 (0.1% titration in mannitol)	2.9736	0.029736 = 0.0000297 g vitamin B12
Calcium pantothenate	6.6079	0.066079
Choline dihydrogen citrate	349.6916	3.496916 = 1.4337 g choline
Folic acid	0.1982	0.001982
Inositol	11.0132	0.110132
Menadione (vitamin K3)	4.9559	0.049559
Niacin	9.9119	0.099119
Pyridoxine HCl	2.2026	0.022026
Riboflavin	2.2026	0.022026
Thiamin HCl	2.2026	0.022026
Dry vitamin A palmitate (500,000 U/g)	3.9648	0.039648 = 19,824 units
Dry vitamin D2 (500,000 U/g)	0.4405	0.004405 2,202.5 units
Dry vitamin E acetate (500 U/g)	24.2291	0.242291 121.15 units
Cornstarch	466.6878	4.666878

Table III. Mineral Mix, Ca-P Deficient (TEKLAD) (#170915)

Contents	Amount
	g/Kg
Sodium chloride	193.7325
Potassium citrate, monohydrate	575.9615
Potassium sulfate	136.1363
Magnesium oxide	62.8322
Manganous carbonate	9.163
Ferric citrate (16.7% Fe)	15.708
Zinc carbonate	4.1888
Cupric carbonate	0.7854
Potassium iodate	0.0262
Sodium selenite	0.0262
Chromium potassium sulfate	1.4399

Note: When used at the rate of 13.369 g/kg of diet, mineral
elements other than Ca and P will be supplied to the diet at
rates similar to those provided by 35.0 g/kg of #170915.

Table IV. Summary of Effects of Alterations in Dietary Protein and Calcium and Phosphorus Sources in Diets of Young Mice

| Parameter | Protein Source | | Variation | | | |
| | | | Phosphorus Intake | | Calcium Intake | |
	Soy	Egg	High	Low	High	Low
Bonebreaking strength (N)	7.42	8.19	7.40	8.21	8.22	8.21
Femur weight (g)	0.039	0.033	0.032	0.036	0.036	0.032
Femur ash weight (g)	0.023	0.021	0.019	0.024	0.022	0.021
Femur phosphorus (mg)	77.8	64.3	68.0	74.2	76.8	65.5
Femur calcium (mg)	39.2	36.0	35.8	39.5	40.2	35.0
Blood serum phosphorus (mg/dl)	5.5	5.7	5.3	5.9	5.8	5.3
Blood serum calcium (mg/dl)	6.9	8.5	7.0	7.0	7.3	8.1
Fecal calcium (mg)	45.8	46.5	42.0	47.0	76.5	16.0
Fecal phosphorus (mg)	161.0	230.0	220.0	97.0	216.0	101.0
Urinary calcium (mg)	0.25	0.26	2.23	0.28	0.25	0.26
Urinary phosphorus (mg)	71.0	73.0	117.0	27.0	54.0	89.0
Weight gain (g)	10.9	10.6	11.4	10.6	11.6	10.1
Feed intake, (g)	165.0	183.0	170.0	178.0	182.0	166.0
FER (g)	0.066	0.059	0.067	0.060	0.064	0.061

0.033 g). Low phosphorus rations resulted in higher femur bone weights than did high phosphorus rations (0.036 g versus 0.032 g); but high calcium rations resulted in higher femur weight than did low calcium diets (0.036 g versus 0.032 g) (Table IV). While high phosphorus diets tended to reduce femur bone weights in both soy and egg white fed mice, the degree of effect was greater in the soy fed mice. Increasing dietary calcium had a greater effect in decreasing the adverse effects of increased dietary phosphorus in soy based rations than in egg white based rations as illustrated in Figure 2.

As would be expected, femur bone ash weight, femur bone calcium content and femur bone phosphorus content, as illustrated in Figures 3 through 5, tended to be reflections of total femur weight, but these relationships were not perfect when applied to specific experimental ration-fed mouse groups. Femur bone ash contents were significantly higher for animals fed the soy rations plus 0.3% calcium and 0.3% phosphorus (0.026 g) or 1.2% calcium and 0.3% phosphorus (0.028 g) and for animals fed the egg white protein ration plus 1.2% calcium and 0.3% phosphorus (0.023 g) than for animals fed other rations. Orthogonal contrast analyses indicated that increased phosphorus levels had a negative effect on ash weight (0.019 versus 0.024, $P < 0.0001$), increased calcium level had a positive effect on bone ash weight (0.022 versus 0.021, $P < 0.0001$), and that use of soy protein rather than egg white protein increased femur ash weight (0.023 versus 0.021, $P < 0.0001$). Phosphorus level x calcium x protein source was also found to influence ash weight ($P < 0.0002$) in the same way as this variable influenced total femur weight.

Significantly more phosphorus was excreted in the feces when 1.2% calcium and 1.2% phosphorus diets were fed with either egg white (317 mg) or soy (303 mg), than when any of the other rations were fed. Orthogonal contrast analysis indicated that statistically significant sources of these differences included ration phosphorus level ($P < 0.001$), calcium level ($P < 0.0001$), and calcium level x protein source ($P < 0.0461$). The degree of effectiveness of increasing phosphorus excretion with increased ration level of calcium was greater in egg white fed animals than in soy fed animals.

Mean fecal calcium loss at high levels of calcium intake was 75.0 mg and at low calcium intake was 16.0 mg. Mean calcium and phosphorus levels are illustrated in Figure 6. Orthogonal contrast analysis indicated that the level of calcium in the ration was the only significant source of difference in the data.

Urinary calcium excretion levels were not significantly affected by the dietary alterations used in this study. However, urinary phosphorus excretion was significantly affected by dietary phosphorus level, dietary calcium level and protein source. Highest urinary phosphorus losses were observed when mice were fed 1.2% calcium and 1.2% phosphorus with either egg white or soy, while the lowest level was observed when mice were fed 1.2% calcium and 0.3% phosphorus with egg white. Mice fed high phosphorus diets had higher urinary phosphorus losses than did those fed low phosphorus diets (117 versus 27 mg/day). The distinct lowering effect on urinary phosphorus loss of increasing dietary calcium was absent primarily at the high level of phosphorus intake. In addition, the degree of depression in urinary phosphorus loss apparently caused by increased calcium intake was greater in degree in the mice fed egg white diets than in those fed soy as illustrated in Figure 7.

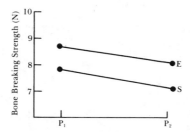

Figure 1. The effect of different ration levels of phosphorus (P_1 = 0.3% and P_2 = 1.2%) and source of protein (E = egg white and S = soy) on bone breaking strength. There was no interaction; main effect was significant. Phosphorus $P < 0.018$; protein $P < 0.027$.

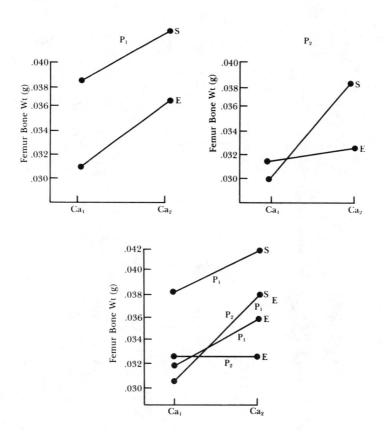

Figure 2. The effect of different ration levels of calcium (Ca_1 = 0.3% and Ca_2 = 1.2%) and phosphorus (P_1 = 0.3% and P_2 = 1.2%) on mice femur bone weight. P–Ca–protein interaction $P < 0.0105$.

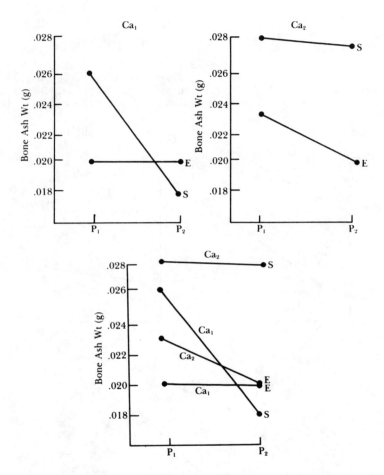

Figure 3. The effect of different ration levels of calcium (Ca$_1$ = 0.3% and Ca$_2$ = 1.2%) and phosphorus (P$_1$ = 0.3% and P$_2$ = 1.2%) on mice femur bone ash weight. P-Ca-protein interaction P < 0.0005.

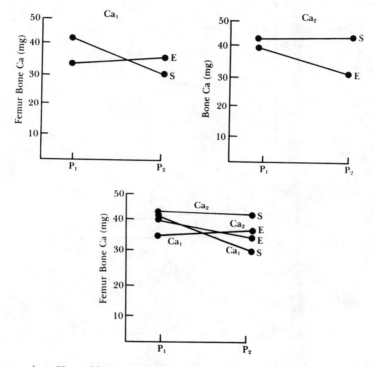

Figure 4.   The effect of different ration levels of calcium (Ca_1 = 0.3% and Ca_2 = 1.2%) and phosphorus (P_1 = 0.3% and P_2 = 1.2%) on mice femur bone calcium.   P–Ca–protein interaction $P < 0.0001$.

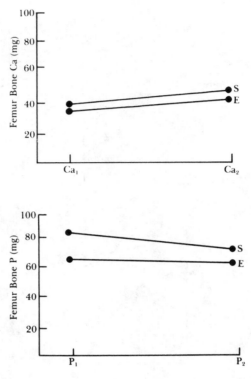

Figure 5.   The effect of different ration levels of calcium (Ca$_1$ = 0.3% and Ca$_2$ = 1.2%) and phosphorus   (P$_1$ = 0.3% and P$_2$ = 1.2%) on mice femur bone phosphorus and calcium when the mice were fed soy (S) or egg white (E) protein.   Top: The main effect, ca. P < 0.0001; no interaction.   Bottom: P-protein interaction P < 0.016.

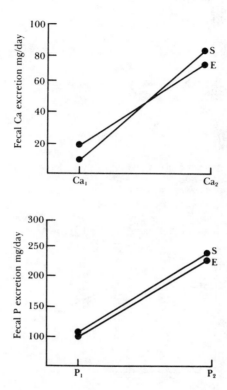

Figure 6. The effect of different ration levels of calcium (Ca$_1$ = 0.3% and Ca$_2$ = 1.2%) and phosphorus (P$_1$ = 0.3% and P$_2$ = 1.2%) on fecal calcium and phosphorus in mice fed soy (S) or egg white (E) diets. Top: Main effect, ca. P < 0.0001. Bottom: P–protein interaction P < 0.0001.

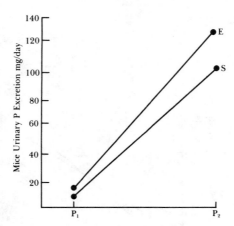

Figure 7. The effect of different ration levels of calcium (Ca_1 = 0.3% and Ca_2 = 1.2%) and phosphorus (P_1 = 0.3% and P_2 = 1.2%) on urinary phosphorus excretion of mice fed soy (S) or egg white (E) protein. P-protein interaction $P < 0.0039$.

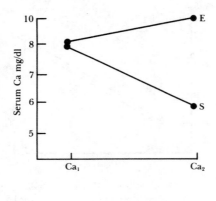

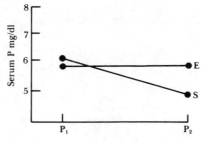

Figure 8. The effect of different ration levels of calcium (Ca_1 = 0.3% and Ca_2 = 1.2%) and phosphorus (P_1 = 0.3% and P_2 = 1.2%) on mice serum levels. Top: Ca-protein interaction $P < 0.0007$. Bottom: Ca-protein interaction $P < 0.0197$.

Blood serum phosphorus levels were significantly lower when the 0.3% calcium and 1.2% phosphorus levels were fed with either soy protein (4.0 mg/dl) or egg white (5.0 mg/dl) than when any of the other test rations were fed. Blood serum phosphorus levels of the mice were higher on the low phosphorus diets than on the high phosphorus diets (5.9 mg/dl versus 5.3 mg/dl). High levels of ration calcium resulted in higher blood serum phosphorus than did lower levels (5.8 versus 5.3 mg/dl).

Figure 8 illustrates the degree of response in serum calcium and phosphorus to the dietary supplementation. Mean blood serum calcium of mice fed the soy plus 1.2% calcium and 1.2% phosphorus or 1.2% calcium and 0.3% phosphorus were significantly lower than were values for all other groups of mice. Egg white fed mice had higher calcium blood serum levels than did mice fed the soy diets (8.5 versus 6.9 mg/dl). While increased dietary calcium had a strong depressing effect when the soy diets were fed, the increased intake of calcium had a slightly positive effect when the egg white diets were fed.

High intakes of protein, particularly animal protein coupled with low phosphorus intakes have been credited as resulting in high urinary losses of calcium and poor calcium balances which can be overcome by increasing intakes of either calcium or phosphorus or lowering protein intake in the adult human (9,15,16). In the present study, not only did the animal protein diet result in bones more resistant to breakage than did the plant protein diet, but the animal protein fed animals were seemingly less affected by variations in intake of phosphorus and calcium than were the plant protein fed animals.

Acknowledgments

Paper published as Nebraska Agriculture Experiment Station Journal Series Number 7623. Funding for this project was provided by Nebraska Agriculture Experiment Station Project 91-024 and by USDA, CSRS Project W-143. Parts of data reported in this paper were taken from N. Nnakwe's Ph.D. dissertation.

Literature Cited

1. Albanese, A.A. In "Current Topics in Nutrition and Disease"; Albanese, A.A. and Kritchevsky, D., Ed.; Alan R. Liss: New York, 1977; Vol. 1, p. 11.
2. Krishnarao, G.V.G. and Draper, H.H. J. Nutr. 1972, 102, 1143.
3. Shah, B., Krishnarao, G. and Draper, H. J. Nutr. 1967, 92, 30.
4. Bauer, K.D. and Griminger, P. J. Nutr. 1983, 113, 2111.
5. Bauer, K.D. and Griminger, P. Nutr. Reports Int. 1984, 29, 1029.
6. Schaafsma, G. and Vissar, R. J. Nutr. 1980, 110, 1101.
7. Sie, T., Draper, H.H. and Bell, R.R. J. Nutr. 1974, 104, 1195.
8. Johnson, N.E., Alcantara, E.N. and Linkswiler, H. J. Nutr. 1970, 100, 1425.
9. Spencer, H., Kramer, L., Osis, D. and Norris, C. J. Nutr. 1978, 108, 447.
10. Linkswiler, H.M., Joyce, C.L. and Anand, C.R. Trans N.Y. Acad. Sci. (Series 11), 1974, 36, 340.

11. Jowsey, J. "Metabolic Disease of Bone"; Saunders Co.:
     Philadelphia, PA, 1977.
12. Herbert, L.A., Lemann, J., Petersen, J.R. and Lennon, E.J. J.
     Clin. Invest. 1966, 45, 1886.
13. Jowsey, J. and Balasubramanian, P. Clin. Sci. 1972, 42, 289.
14. Fiske, C.H. and Subbarow, J. J. Biol. Chem. 1925, 66, 374.
15. Nnakwe, N. Ph.D. Dissertation, University of Nebraska, 1984.
16. Atinmo, T., Taylor, G.O., Mbofung, C.M.F., Fajim, J.L.,
     Adenaike, F.A. Nutr. Res. 1982, 2, 569.

RECEIVED December 26, 1984

Effect of Oxalic Acid on Calcium Bioavailability

JUNE L. KELSAY

Carbohydrate Nutrition Laboratory, Beltsville Human Nutrition Research Center, Agricultural Research Service, U.S. Department of Agriculture, Beltsville, MD 20705

Review of studies on the effect of oxalic acid on calcium bioavailability in rats and in humans indicates that most of the research was done between 1930 and 1950. Decreased availability of calcium in young rats was reported when spinach containing oxalic acid was fed with low calcium diets. The extent of the effect of oxalic acid on calcium availability was shown to be related to levels of calcium and oxalic acid, as well as the presence of vitamin D in the diet. In human studies there was generally no effect of oxalic acid on calcium balance; however, in a few studies decreased calcium balances were reported. There is recent evidence that oxalic acid consumed along with a moderately high level of fiber intake may have adverse effects on calcium balance of human subjects.

Decreased calcium balances in children fed vegetables were reported by Sherman and Hawley (1) in 1922. The vegetables fed were carrots and spinach, and the decreased balances were at that time attributed to the crude fiber in the vegetables. However, results of experiments conducted on rats in the 1930's led investigators to believe that the consumption of spinach as a source of calcium was undesirable due to the presence of oxalic acid in large amounts. Oxalic acid can combine with calcium to form a salt complex of low solubility. Spinach has been reported to contain oxalic acid in amounts from 458-1050 mg/100 g (2-6), depending on the variety of spinach. Similar amounts of oxalic acid are found in other green vegetables such as beet leaves, Swiss chard, lamb's quarter, purslane, dandelions, and poke. Rhubarb, cashew nuts, almonds, walnuts, lady finger, cocoa, and dried tea leaves also contain considerable amounts of oxalic acid.
  Several comprehensive reviews on oxalic acid have been published in which effects on calcium metabolism were discussed (7-11). A review of studies on the effect of oxalic acid on calcium bioavailability in rats and humans is presented in this paper.

Studies on Rats

Bloom (12) reported that in both older rats (63 days of age) and younger rats (33 days of age) the amount of calcium retained was less when 5% dried spinach, either raw or cooked, was included in the diet in 1-week balance studies (Figure 1). The experimental diets contained about 0.4% calcium, and one drop of cod liver oil was given each rat every day. Most of the calcium excretion on the spinach diet was in the feces. Retentions of calcium on the basal low fiber diet and on diets containing filter paper in amounts to equal the crude fiber in spinach or 12 times the crude fiber in spinach were high. There was no significant difference in calcium retentions between raw and cooked spinach. The low retention of calcium from spinach could not be attributed to the presence of crude fiber in the diet.

In a study by Fincke and Sherman (13), the calcium of spinach was not utilized as well as that from milk; however, the calcium of kale, which is low in oxalic acid (3,4), was about as available as that from milk. The calcium utilization factor was determined by dividing the weight of calcium stored by the weight of calcium ingestion. Rats 4 weeks old were fed for 60 days a diet in which most of the calcium was supplied by skim milk, or in which half of the skim milk was replaced by dried spinach or dried kale in amounts to provide the same amount of calcium. The diets contained about 0.3% calcium and 10% butter fat. It was concluded that the poor utilization of the calcium of spinach was due to the oxalic acid in spinach.

Tisdall et al. (14) obtained poor weight gains and low calcium retention in rats fed 10% dried, cooked spinach with a low calcium (0.041 g percent) diet. Rats 28 days of age were placed on a low calcium diet, or the low calcium diet containing 10% dried spinach, 5.83% dried tomatoes, or 1.88% dried spinach for 4 weeks each. Calcium contents of whole rats were determined at the end of the experiment. Weight gains and calcium retentions were better when the animals were fed tomatoes than when they were fed spinach, even though the calcium intake on the higher level of spinach was about 20 times greater than when tomatoes were fed.

The results of experiments conducted by MacKenzie and McCollum (15) indicate that the effect of dietary oxalic acid on the rat depends on the composition of the diet. There was no effect on rate of growth or calcium excretion of 50 g rats fed for 10 weeks a diet containing 0.6% calcium, 0.7% phosphorus, and optimum vitamin D, when levels of potassium oxalate up to 2.5% were fed. The percent bone ash on the 2.5% oxalate diet was somewhat lower than on the control diet. On a 0.35% calcium, 0.35% phosphorus, and vitamin D-free diet, 1.7% potassium oxalate resulted in restricted growth and bone formation of weanling rats.

Tisdall and Drake (16) found that the addition of 10% dried spinach to a low calcium diet containing 2% cod liver oil resulted in a reduction in body calcium (Figure 2). The spinach diet was compared with one containing an equal amount of calcium as calcium carbonate, and with a third containing an equal amount of calcium carbonate plus an amount of oxalic acid equal to that present in the spinach. Three-week-old rats were fed the diets for 4 weeks.

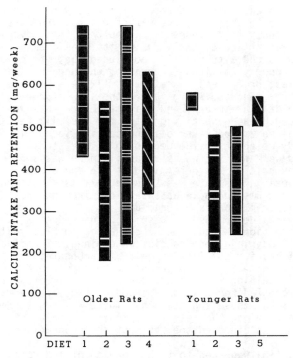

Figure 1.   Calcium intake and excretion of older rats (63 days) and younger rats (33 days) on a low fiber diet (1), raw spinach (2), cooked spinach (3), filter paper (4), and high filter paper (5).   Dietary intake is indicated by the top of the bar.   Balance is indicated by the bottom of the bar.   Adapted from Ref. 12.

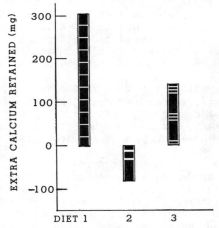

Figure 2.   Extra calcium retained by feeding rats added calcium as 0.115% calcium carbonate (1), as spinach (2), and as calcium carbonate + oxalic acid (3).   Adapted from Ref. 16.

Although there was decreased retention of body calcium when oxalic acid was added to the diet, the effect was not as drastic as when spinach was fed. The authors concluded that this adverse effect of spinach was at least partly due to its oxalic acid content.

Fairbanks and Mitchell (17) concluded that the calcium of spinach was poorly utilized for growth, whether the spinach was fresh, cooked, or canned. Rats given milk were pair-fed to those given spinach. Sucrose was given to the slower-growing rat of each pair. Calcium intakes and weights of the rats were equalized, but body length was generally greater for the milk-fed rats. The diets contained 2% cod liver oil and calcium to supply 5 mg per g gain in body weight. At the beginning of the study the rats weighed 30-70 g and were fed 5-6 weeks or until there was a gain of about 100 g. Body calcium was determined at the end of the study. The calcium of calcium oxalate was considerably less utilizable than that from calcium gluconate as determined from balances on rats receiving liberal amounts of calcium.

Kohman (3) determined that spinach interfered with growth and bone formation. Studies were conducted on 21-day-old rats for 21 or 28 days and body calcium determined. The diets furnished about 0.22% calcium supplied by calcium carbonate, calcium oxalate, 8.2% spinach, or 4.5% turnip greens. The supplements were added to a diet containing meat, peas, carrots, and sweet potatoes. When rats were fed for a longer period of time, there was a high death rate among those fed spinach. There was considerable oxalic acid in the feces of the rats fed spinach. Results of these studies could not be explained on the basis of the crude fiber content of spinach.

Spiers (18) reported that the feeding of 14% dried New England spinach in the diet resulted in decreased growth and less calcium retention, as determined by body calcium content (Figure 3). Rats 28 days old were fed until 60 days of age. All diets contained about 0.3% calcium, and in the control diet practically all of the calcium was from skim milk. In the other diets, one-half of the skim milk was replaced by turnip greens, tendergreens, collards, kale, or New Zealand spinach to furnish an equal amount of calcium. Spinach was the only one of the greens fed which contained considerable oxalic acid. The calcium of the spinach appeared not only to be utilized poorly, but spinach also impaired utilization of the milk calcium. Food intake of the rats on the spinach diet was also lower.

Fitzhugh and Nelson (19) found that oxalic acid up to 1.2% of the diet did not affect growth or mortality rate of rats fed oxalic acid for 1 year. Microscopic pathological examination showed no major visceral damage, but some of the rats showed slight periportal hypertrophy of the hepatic cells along with slight centrolobular atrophy.

Lovelace et al. (20) noted that the effect of oxalic acid on calcium retention depended on the age of the animal (Figure 4). The two experimental diets contained cod liver oil and 0.3% calcium. Milk was the main source of calcium, and to one diet was added 60 mg of potassium oxalate daily. Ten-day balances were determined throughout the study until rats were 175 days of age. At 50 days of age, rats fed oxalate retained about 25% of the calcium,

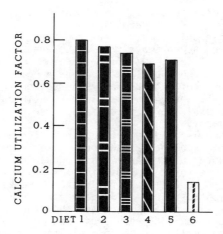

Figure 3. Calcium utilization factor of rats fed control diet
(1), turnip greens (2), tendergreens (3), collards (4), kale
(5), and New Zealand spinach (6). Adapted from Ref. 18.

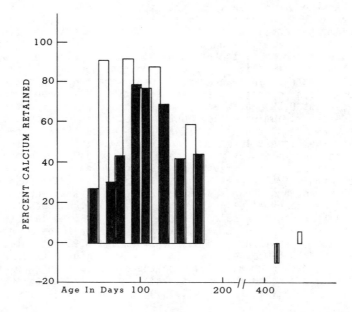

Figure 4. Percent calcium retained at different age levels by rats
fed milk with oxalate and milk without oxalate. Key: ■, milk
with oxalate; and □, milk without oxalate. Adapted from Ref. 20.

and rats fed no oxalate retained about 90%. As the rats fed oxalate
grew older, calcium retention increased until about 100 days.
At this point they retained almost 80% of the calcium as compared
with 88% retained by those not receiving oxalate. Retention of
calcium declined after this time, and at 150 days of age rats fed
oxalate were rataining approximately 40%, whereas those fed no
oxalate were retaining almost 60% of the calcium. From 175-438
days of age, rats were fed a stock diet containing 3.3% calcium.
At 428 days the experimental diets were resumed for two 14-day
periods. When the rats were 450 days of age, their calcium retention
was about 6% when no oxalate was added, and those given oxalate
were in negative balance.

In a study by Mueller and Cooney (21) rats fed a diet containing
0.45% calcium and 16% cocoa powder showed a lower growth rate and
body calcium than those on a control diet. However, oxalic acid
did not account for the total decrease in body calcium. Rats 25-
27 days old were pair-fed the basal ration for 5 weeks. The only
variable in the ration was cocoa. The decreased growth rate occurred
in the cocoa-fed rats in spite of the fact that they received more
calories, protein, calcium and phosphorus than controls.

Newell and Elvehjem (22) found no inhibition of growth in
21-day-old rats fed a diet containing 86.3 parts of whole milk
and 11.7 parts of chocolate syrup for 16 weeks. Cocoa furnished
1.05% of the diet, which equals about 7-8% cocoa on the dry basis.
The authors commented that this amount represented a more likely
intake than the amount fed in the study by Mueller and Cooney (21).

In a study in which Mitchell and Hamilton (23) duplicated
the experimental conditions of Mueller and Cooney (21), calcium
retention of rats was less when cocoa was fed. The investigators
incorporated the cocoa into the basal diet at the expense of sugar.
Body weights were equalized by adding sugar to the diet of slow-
growing pair-mates. Three brands of cocoa were fed, and one brand
resulted in less retention than the other two. All three brands
contained about 0.5% oxalic acid, but the brand which resulted
in the lowest calcium retention contained 5.2% crude fiber and
the other two brands contained 3.6 and 3.8% crude fiber. The cocoa
containing the highest level of crude fiber was the same cocoa
used in the experiments by Mueller and Cooney (21). Therefore,
the added effect of fiber may be of some consequence.

Tuba et al. (24) reported that rats fed 15.4% dried rhubarb
or an equivalent amount of sodium oxalate grew at a slower rate
and had higher serum alkaline phosphatase levels than rats fed
control diets containing calcium lactate. The calcium lactate
diets contained either 0.5% or 0.09% calcium. The rhubarb and
sodium oxalate diets contained 0.5% calcium and 0.95% oxalate.
Diets were fed to 6-week-old rats for 6 weeks. Ash content of
tibia and serum calcium were lower on the diets containing oxalate.
Four out of six rats on the sodium oxalate diet died after the
fourth week, and all of the rats lost weight on this diet.

Studies on Humans

Sherman and Hawley (1) studied the effects of replacing milk with
carrots and spinach in the diets of three children of 6, 10, and

12 years of age. The children were given 750 g of milk/day for
9 days. The milk was then reduced to half and carrots and spinach
added to replace the other half of the calcium. Calcium intakes
were 0.894 to 1.044 g/day and spinach intakes were 196-350 g/day.
Mean 3-day calcium balances, determined for from 2-5 periods, were
positive for all diets, but were lower on the diets containing
carrots and spinach. The authors concluded that children do not
utilize calcium from vegetables as well as that from milk.

   McLaughlin (25) reported that although calcium balances for
seven women were somewhat lower during 6 days in which spinach
replaced milk in the diet, all balances were positive. The women
were fed diets containing about 500 mg of calcium/day in which
79% came from milk for 6 days and 73% came from spinach (about
276 g/day) for 6 days. The spinach diet contained 2.0 g oxalic
acid/day. The calcium excretion in urine was 2-3 times greater
during the milk period.

   Schultz et al. (26) found that feeding spinach to four infants
for 3-6 days did not affect calcium balance, with the possible
exception of a slight lowering when dried spinach was fed. Infants
5 weeks-6 months of age were fed only milk formula for a 6-day
control period. The spinach was added to the formula diet in the
following amounts: 6 g dried spinach, 60 g pureed spinach, or
70 g raw spinach per day. The infants were also given cellulose
in amounts of 6 or 9 g/day. There was generally a larger output
of feces when the diet contained spinach.

   In a study by Bonner et al. (27) on 10 children 5-8 years
of age, cumulative calcium balances showed that neither spinach
nor oxalic acid changed the rate of storage of calcium or showed
a deleterious after-effect during the subsequent control period
(Figure 5). There was an initial period of 25 days on the basal
diet, followed by 15 days on the control diet supplemented with
100 g spinach per day, then 5 days on the same diet in which oxalic
acid and calcium substituted for their equivalents in 100 g spinach,
and finally another 15 days on the basal diet. Five-day balances
were determined throughout the study. Daily calcium intake was
from 0.733-1.324 g/day, and the spinach contained 0.7 g oxalic
acid/100 g. Storage rates of nitrogen, calcium, and phorphorus
were not significantly affected by the addition of spinach or oxalic
acid to the diet.

   In two women studied by Fincke and Garrison (28), the calcium
of kale was better utilized than that of spinach, when the vegetables
supplied 80-86% of the calcium in the diet. One diet contained
kale and two diets contained spinach (about 500 g); calcium intakes
were 400, 304, and 445 mg/day, respectively. Two 3-day balances
were determined after 3 days on each diet. Calcium balances were
negative on all diets, but were more negative when the spinach
was consumed.

   Six women studied by Johnston et al. (29) did not have significantly different calcium balances when 120 g spinach/day was added
to their diet (Figure 6). The women were fed a basal diet for
4 weeks, the basal diet plus 120 g spinach for breakfast for 4
weeks, and the basal diet plus 120 g spinach for dinner for 4 weeks.
Half the women consumed spinach with breakfast and half with dinner
during the second 4 weeks; during the third 4 weeks the order was

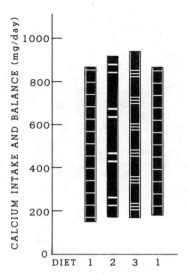

Figure 5. Mean calcium intake and balance of 10 children consuming a control diet (1), spinach (2), and oxalic acid (3). Dietary intake is indicated by the top of the bar. Balance is indicated by the bottom of the bar. Adapted from Ref. 27.

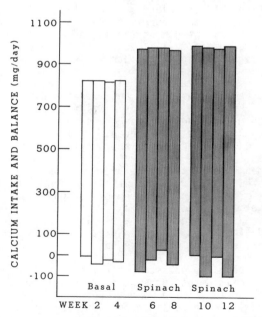

Figure 6. Mean calcium intake and balance of six women consuming a basal diet and the basal diet containing spinach. Dietary intake is indicated by the top of the bar. Balance is indicated by the bottom of the bar. Adapted from Ref. 29.

reversed. All milk and cheese was given with the noon meal. Calcium intakes were 820 mg/day on the basal diet, and 975 and 982 mg/day on the diets containing spinach. More calcium was excreted in the feces when spinach was added to the diet, but the difference was not significant. The oxalic acid appeared to prevent the absorption of calcium in spinach.

Bricker et al. (30) reported that there were no statistically significant differences between the calcium balances of eight women on cocoa and non-cocoa diets. The women were studied for three to seven 4-day periods. Calcium intake was 670 mg/day with the addition of milk and 679 and 755 mg/day with the addition of milk and cocoa. Five levels of cocoa, supplying from 5.6-52.6 g/day, were tested. These amounts would likely contain from 25-280 mg of oxalic acid, which was not nearly as much as was added when spinach was fed. With the inclusion of cocoa in the diet, the urinary calcium fell and fecal calcium rose. There were also increases in the fecal excretion of dry matter and nitrogen.

In a study conducted at the Beltsville Human Nutrition Research Center (31), we found a decrease in calcium balance when spinach was fed in a moderately high fiber diet for 4 weeks (Figure 7). Twelve men consumed three diets for 4 weeks each. Diet 1, a low fiber diet, contained 5 g neutral detergent fiber/day (NDF) and about 100 g spinach every other day. Diet 2 contained 27 g NDF/day in fruits and vegetables, and about 100 g spinach every other day. Diet 3 was the same as diet 2 except that cauliflower, which is low in oxalic acid, replaced the spinach. There were no significant differences in calcium balances among the three diets during week 3, but during week 4 the calcium balance on the higher fiber diet containing spinach was significantly lower than that on the low fiber diet containing spinach. These results indicate that there is a combined effect of oxalic acid and fiber on the availability of calcium and suggest that the effect is also time related.

In studies on test meals, Walker et al. (32) discovered that the calcium of Swiss chard, which has a high oxalic acid content, was poorly absorbed. Children excreted more calcium during the 6 hours after a test meal of milk, Pumpkin leaves, cassava leaves, or pigweed leaves than after Swiss chard. All supplements contained 200 mg calcium.

Pingle and Ramasastri (33) also found that the availability of calcium from amaranthus, which contains oxalic acid, was low compared to that of milk (Figure 8). Amaranthus given along with milk also decreased the availability of calcium from milk. Urinary calcium excretion of adults was determined for 6 hours following test meals. Pingle and Ramasastri (34) further reported that when the cooking water was drained from the amaranthus, calcium absorption was higher than when the cooking water was included, indicating that the soluble oxalates were removed in the cooking water.

Summary and Conclusions

In the studies on rats, the amounts of spinach, cocoa, rhubarb, or oxalate salts fed were much higher than would normally be included in diets. The levels of calcium and vitamin D were generally low. For maximum calcification during growth, calcium should be 0.5%

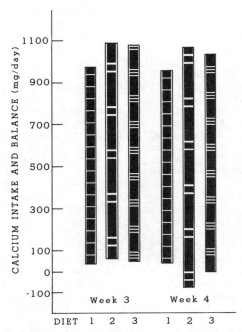

Figure 7. Mean calcium intake and balance of 12 men consuming
a low fiber diet containing spinach (1), a diet containing fruits
and vegetables with spinach (2), and a diet containing fruits
and vegetables in which cauliflower replaced spinach (3). Dietary
intake is indicated by the top of the bar. Balance is indicated
by the bottom of the bar. Adapted from Ref. 31.

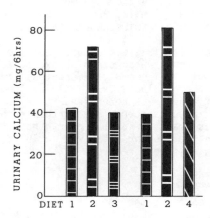

Figure 8. Urinary calcium of male adults consuming a basal diet
(1), milk diet (2), amaranthus diet (3), and a milk + amaranthus
diet (4). Adapted from Ref. 33.

of the diet, and 1000 I.U. of vitamin D per kg of diet is recommended
(35). Decreased growth rates and decreased body calcium observed
in these studies were greatest when high levels of oxalic acid
and low levels of calcium were fed. The effect of oxalic acid
was overcome when the level of calcium was sufficient. Ackerman
and Gebauer (36) carried out experiments on rats which showed that
there was satisfactory growth when the molar ratio of calcium:oxalate
was 1.0:0.5, but that molar amounts of oxalic acid exceeding those
of calcium in the diet could lead to death.

In the studies on humans there appeared to be decreased calcium
balances when 200 g or more of spinach per day was included in
the diet. In two of the studies in which women were fed spinach,
calcium intakes were below the Recommended Dietary Allowance of
800 mg/day (37). Some studies were conducted for short period
of a week or less, which may not be sufficient time to adjust to
a change in diet. From measurement of calcium excretion in urine
after a test meal, it was shown that the calcium in oxalate-contain-
ing vegetables was less well-absorbed than that of milk or of vegetables not containing oxalic acid. However, this would not necessarily affect calcium balance, since the total amount of calcium
in the diet would have to be considered. The effect of a combination
of oxalic acid and fiber on calcium bioavailability should be further
investigated.

Literature Cited

1. Sherman, H. C.; Hawley, E. J. Biol. Chem. 1922, 53, 375-99.
2. Majumdar, B. N.; De, N. K. Ind. J. Med. Res. 1938, 25, 671-5.
3. Kohman, E. F. J. Nutr. 1939, 18, 233-46.
4. Andrews, J. C.; Viser, E. T. Food Res. 1951, 16, 306-12.
5. Zarembski, P. M.; Hodgkinson, A. Brit. J. Nutr. 1962, 16, 627-34.
6. Kaul, S.; Verma, S. L. Ind. J. Med. Res. 1967, 55, 274-8.
7. Jeghers, H.; Murphy, R. New Eng. J. Med. 1945, 233, 208-15.
8. Fassett, D. W. In "Toxicants Occurring Naturally in Foods"; 2nd ed. National Academy of Sciences: Washington, D. C., 1973; pp. 346-62.
9. Gontzea, I.; Sutzescu, P. In "Natural Antinutritive Substances in Foodstuffs and Forages"; S. Karger: Basel/New York, 1968; pp. 84-108.
10. Oke, O. L. World Rev. Nutr. Dietet. 1969, 10, 262-303.
11. Hodgkinson, A. "Oxalic Acid in Biology and Medicine": Academic Press: London, 1977.
12. Bloom, M. A. J. Biol. Chem. 1930, 89, 221-33.
13. Fincke, M. L.; Sherman, H. C. J. Biol Chem. 1935, 110, 421-8.
14. Tisdall, F. F.; Drake, T. G. H.; Summerfeldt, P.; Jackson, S. H. J. Pediat. 1937, 11, 374-84.
15. MacKenzie, C. G.; McCollum, E. V. Am. J. Hygiene 1937, 25, 1-10.
16. Tisdall, F. F.; Drake, T. G. H. J. Nutr. 1938, 16, 613-20.
17. Fairbanks, B. W.; Mitchell, H. H. J. Nutr. 1938, 16, 79-89.
18. Spiers, M. J. Nutr. 1939, 17, 557-64.
19. Fitzhugh, O. G.; Nelson, A. A. J. Am. Pharm. Assoc. Sci. Ed. 1947, 36, 217-19.

20. Lovelace, F. E.; Liu, C. H.; McKay, C. M. Arch. Biochem.
     Biophys. 1950, 27, 48-56.
21. Mueller, W. S.; Cooney, M. R. J. Dairy Sci. 1943, 26, 951-8.
22. Newell, G. W.; Elvehjem, C. A. Science 1944, 99, 411-2.
23. Mitchell, H. H.; Hamilton, T. S. J. Nutr. 1946, 31, 377-86.
24. Tuba, J.; Siluch, K. A.; Robinson, M. I.; Madsen, N. B. Canad.
     J. Med. Sci. 1952, 30, 515-9.
25. McLauglhlin, L. J. Biol. Chem. 1927, 74, 455-62.
26. Schlutz, F. W.; Morse, M.; Oldham, H. Am. J. Dis. Child.
     1933, 46, 757-74.
27. Bonner, P.; Hummel, F.C.; Bates, M. F.; Horton, J.; Hunscher,
     H. A.; Macy, I.G. J. Pediat. 1938, 12, 188-96.
28. Fincke, M. L.; Garrison, E. A. Food Res. 1938, 3, 575-81.
29. Johnston, F. A.; McMillan, T. J.; Falconer, G. D. J. Am.
     Dietet. Assoc. 1952, 28, 933-8.
30. Bricker, M. L.; Smith, J. M.; Hamilton, T. S.; Mitchell, H. H.
     J. Nutr. 1949, 39, 445-62.
31. Kelsay, J. L.; Prather, E. S. Am. J. Clin. Nutr. 1983, 38,
     12-9.
32. Walker, A. R. P.; Walker, B. F.; Wadvalla, M. Ecol. Fd. Nutr.
     1975, 4, 125-30.
33. Pingle, U.; Ramasastri, B. V. Br. J. Nutr. 1978, 39 119-25.
34. Pingle, U.; Ramasastri, B. V. Br. J. Nutr. 1978, 40, 591-4.
35. "Nutrient Requirements of Laboratory Animals," National Academy
     of Sciences, 1978 3rd ed.
36. Ackermann, H.; Gebauer, H. Nahrung 1958, 12, 278-5.
37. "Recommended Dietary Allowances," National Academy of Sciences,
     1980, 9th ed.

RECEIVED January 19, 1985

Evidence for a Role for Insulin in the Hypercalciuretic Response to Macronutrients

LINDSAY H. ALLEN

Department of Nutritional Sciences, University of Connecticut, Storrs, CT 06268

During the last decade, a number of investigators have attempted to elucidate the mechanism by which protein consumption induces hypercalciuria. In our previous research, the calciuria could not be explained by increased intestinal absorption of calcium, complexing with sulfur amino acids, or urinary acid production. This paper summarizes our evidence that the mechanism involves protein stimulation of insulin secretion, followed rapidly by insulin inhibition of renal calcium reabsorption. In humans, urinary calcium was proportional to peak postprandial insulin levels in several experiments, after either protein or sucrose was fed. In rats, infusion of the insulin secretagogues arginine and glucose induced a calciuria proportional to serum insulin levels. Suppression of insulin secretion by mannoheptulose or streptozotocin prevented the calciuria. Parathyroidectomy did not affect arginine-induced hypercalciuria in the rat, so insulin is not inhibiting parathyroid hormone secretion or activity. From preliminary in vitro experiments using rat renal slices, it appears that the hormone directly affects renal calcium transport. If insulin is the mediator of the hypercalciuria, it might be possible to reduce urinary calcium loss by lowering the intake of insulinogenic foods. This would be especially important in those individuals with a marked calciuric response to such foods.

For more than forty years, it has been known that increasing the protein content of the diet causes an increase in urinary calcium excretion (1,2). There is, in fact, a direct correlation between urine calcium output and dietary protein level, so that excretion is 800 percent higher if dietary protein is increased from 6 g per day to 560 g per day (3). This relationship between urinary calcium and protein ingestion is not affected by the level of dietary calcium, and is evident even when severely calcium-deficient diets are consumed (3).

0097–6156/85/0275–0117$06.00/0
© 1985 American Chemical Society

Since the early 1970's, research has been directed at identify-
ing the mechanism by which the calciuria is induced. Attention was
given first to the question of whether the elevated urinary calcium
excretion was caused by an increase in the intestinal absorption of
calcium. Results of calcium balance studies in human subjects showed
that protein ingestion either had no effect on calcium absorption (4)
or that the effect was insufficient to account for the calciuria (5).
Consequently, negative calcium balance is a frequent observation in
human studies when high protein diets are fed, and this situation is
not improved by high calcium intakes (4,5).

The next step toward the elucidation of how the calciuria is induced was to focus on the possibility that protein impaired the renal
reabsorption of calcium. Using postprandial response to single meals
as the experimental approach, we demonstrated that within 30 minutes
after its consumption, a high protein meal significantly impaired
renal calcium reabsorption (6). The peak rate of calcium excretion
occurred at 1.5 hours postprandially, when calcium reabsorption was
lowest. Protein-induced impairment of calcium reabsorption by the
kidney has since been confirmed by Schuette et al. (7). These
authors, in a 30-day balance study, found that glomerular filtration
rate (GFR) was increased by high protein feeding, and that this contributed to the hypercalciuria. However, such changes in GFR only
occur after a relatively long period of continuously consuming a
high protein diet (4,8), as a consequence of hypertrophy and hyper-
plasia of glomeruli, tubules and interstitial cells (9). The fact
that GFR is not increased by feeding a single meal of high protein
content, yet hypercalciuria still occurs, argues against increased
GFR being the primary mechanism responsible for protein-induced
hypercalciuria (6).

It has also been suggested that the sulfur amino acid content of
protein might be responsible for the calciuria, as a consequence of
binding of calcium with sulfate (10), or of the acidosis resulting
from sulfur oxidation (11-13). Using a meal-feeding approach, we
compared the effects in the same subjects of feeding meals containing
15 g protein (control), 45 g protein (high protein), or 15 g protein
plus sulfur amino acids equivalent to those contained in the high
protein meal. Urinary calcium was significantly increased by the
high protein diet, but not by the control diet with added sulfur
amino acids (14). Thus, the sulfur amino acid content of the high
protein diet was not the cause of the hypercalciuria. There was no
difference among the diets in the excretion of net renal acid, urine
pH, titratable acidity, or ammonium ions. An effect of protein
greater than that which could be explained by its constituent amino
acids has been confirmed by others using animals (13) and human
subjects (15).

Evidence for Insulin Involvement in Macronutrient-Induced Hypercalciuria

In recent years, we have accumulated considerable evidence that the
protein-induced impairment of renal calcium reabsorption, and the
subsequent calciuria, are the result of protein-stimulated increases

in insulin secretion. Insulin infusion in euglycemic human subjects
has been demonstrated to increase urinary calcium by impairing the
reabsorption of calcium by the kidney (16). In addition, a search of
the literature revealed that consumption of many carbohydrates produces calciuria (17-20), whereas macronutrients which do not stimulate insulin secretion, such as fat and xylose (18), have no calciuric effect.

Based on this evidence which suggested that insulin may be involved in the calciuric response to protein ingestion, insulin response of human subject was measured during meal-feeding experiments.
Serum insulin was significantly higher within 30 minutes of consuming
a high protein meal compared to a control meal. Of considerable interest was the observation that the magnitude of the calciuric effect
of protein varied widely among individuals (Figure 1) and this variation was correlated with the range in serum insulin response
(r=0.61, P<0.05) (21). Three out of 12 subjects showed no protein-
induced calciuretic response; these same subjects also showed no increase in serum insulin after eating high protein meals compared to
control meals.

We have also investigated the relationship between serum insulin
response to sucrose ingestion and urinary calcium excretion (22). In
a group of 13 healthy adults, we tested whether inter-individual
calciuric responses to the ingestion of sucrose (2 g/kg body weight
consumed as a single drink) were associated with postprandial serum
insulin levels. Urine calcium increased from a fasting level of
50 μg/min to peak at 179 μg/min within 1.5 hours of sucrose ingestion. There was no increase in GFR or filterable serum calcium.
Among individuals, the peak rate of calcium excretion ranged from 119
to 377 μg/min, and urine calcium was negatively correlated with the
percentage of renal calcium reabsorption (r=-0.98). In most subjects, peak levels of serum insulin occurred at 0.5 hours after the
sucrose was consumed, and ranged among subjects from 26 to 123 μU/ml.
As hypothesized, there was a correlation between peak serum insulin
levels and postprandial (three hour) urine calcium output (Figure 2).
As observed in the case of protein feeding, some subjects showed considerable insulin and urinary calcium responses to sucrose ingestion
while in others there was only a small response in both parameters.
It would seem important to establish in future work whether those individuals with the greatest postprandial calciuria are at greater
risk of calcium depletion during the aging process.

Animal Studies

Most conclusive evidence of insulin's role in the inhibition of renal
calcium reabsorption has been obtained from animal studies. Serum
insulin levels were manipulated acutely in rats by infusions of arginine or glucose (23). These nutrients were infused into the jugular
vein, together with ^{3}H-inulin for the measurement of GFR. Urine and
blood samples were collected during a one-hour equilibration period
when saline was infused, and for 1.5 hours after arginine (12 mmol/kg
/hour) or glucose (12 mmol/kg/hour) infusion commenced. Arginine infusion caused a 575 percent increase in urine calcium, while glucose
infusion resulted in a 225 percent increase, compared to saline-
infused controls (Figure 3). Plasma insulin was 200 percent higher

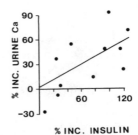

Figure 1. Regression of percent increase in urine calcium against percent increase in serum insulin.

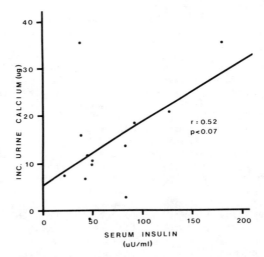

Figure 2. Peak serum insulin levels vs postprandial urine calcium excretion after a sucrose load.

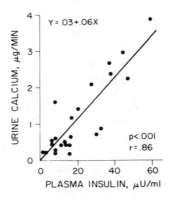

Figure 3. Regression of plasma insulin levels against urinary calcium excretion for animals in all treatment groups.

in the arginine group, and 146 percent higher when infused with glucose.

In additional groups of animals, an injection of mannoheptulose (1 μg/kg) was given at mid-point of the equilibration period, in order to suppress insulin release. Arginine or glucose infusion was commenced 30 minutes later, and urine and plasma samples collected for the next 1.5 hours. Mannoheptulose inhibited the insulinogenic response to arginine by 50 percent, and to glucose by 80 percent. It also severely reduced the calciuric response to these nutrients, by 40 percent for arginine, and by 60 percent in the case of glucose. The regression of plasma insulin levels against urinary calcium excretion for individual animals in all treatment groups was highly significant (r=0.86, P<0.001) (Figure 3). Arginine and glucose infusion stimulated urine flow rate, and glucose increased GFR, but the significant stimulatory effect of these nutrients on urinary calcium excretion persisted when expressed as calcium excreted per ml of glomerular filtrate. The regression between urine calcium per ml glomerular filtrate and plasma insulin levels was r=0.68 (P<0.001).

The importance of insulin as a mediator of the hypercalciuric effect of arginine infusion was also evident from studies conducted in chronically diabetic rats, where diabetes was induced by streptozotocin (23). Animals were injected with streptozotocin prior to arginine infusion; 100 mg/kg i.p. was given on the seventh day before, followed by 25 mg/kg six days before the arginine infusion and renal clearance studies. In contrast to non-diabetic controls, diabetic animals did not increase their urinary calcium excreted (per ml glomerular filtrate) in response to the arginine infusion, nor did the arginine stimulate insulin secretion.

The above experiments strongly suggest to us that a linear relationship exists between serum or plasma insulin levels over a wide physiological range, and urinary calcium excretion. The calciuric response to arginine or glucose infusion does not occur if insulin secretion is prevented, as evidenced by the data obtained from animals made acutely insulinopenic by mannoheptulose, or more chronically diabetic by streptozotocin.

The Mechanism of Insulin Inhibition of Renal Calcium Reabsorption

The mechanism by which insulin impairs renal calcium reabsorption is still under investigation, but some progress has been made toward revealing that the insulin effect is directly on the renal tissues, rather than an indirect effect on other hormones or metabolites.

We have tested the hypothesis that insulin inhibits the stimulatory effect of parathyroid hormone (PTH) on calcium reabsorption in the distal nephron. PTH is known to enhance calcium transport in renal cells, probably by stimulation of adenylate cyclase and subsequent increases in 3'5' cyclic AMP productoin. Since insulin had been observed to inhibit PTH-stimulated increases in kidney cyclic AMP levels in vitro (24) we investigated whether insulin-mediated hypercalciuria was dependent on the presence of PTH in vivo.

The experiment design was to compare the calciuric response to arginine infusion in parathyroidectomized (PTX) and sham-operated (control) rats (25). Four experimental groups were used: PTX infused with saline or arginine (12 mmol/kg/hour), and sham-operated similarly infused with saline or arginine. A one-hour equilibration period of saline infusion was followed by three, 0.5 hour periods of urine and plasma collection. Tritiated insulin was present in all infusates to permit calculation of GFR.

As anticipated, arginine infusion caused a large (221 percent) increase in calcium excretion in sham-operated animals. Parathyroidectomy had no effect on the calciuric response to arginine; uriary calcium increases in PTX control and arginine-infused animals were 299 and 302 percent respectively. These results persisted when data were corrected for differences in GFR. The data illustrate that neither PTH activity nor secretion is involved in insulin impairment of renal calcium transport.

Roy and Seely showed that glucose infusion produces calciuria in both intact and PTX rats (26). These authors suggested that glucose per se, rather than insulin, affects renal calcium transport. However, our previous data showing a lack of glucose effect if insulin secretion is suppressed do not support a direct role for glucose (23). Neither the consumption of a high protein meal (21) nor that of a long-term high protein diet (27) produces changes in serum PTH or urinary cyclic AMP levels. Thus, there is no evidence at present to suggest that the higher insulin levels after the consumption of macronutrients are affecting renal calcium transport via an effect on PTH secretion or activity.

In our recent work we have looked for a direct effect of insulin on calcium transport using kidney slices in vitro (28). This technique avoids the potentially confounding effects of insulin on the levels, secretion or activity of other hormones or metabolites. The procedure is based on that developed by Borle (29). Decapsulated slices of renal cortex were incubated for one hour in the presence of 0, 25, 50 or 100 μU insulin per ml, followed by a second hour of incubation when ^{45}Ca was also added to the slices. At this point, still in the presence or absence of insulin, the radioactive calcium was gradually removed from the slices by eluting them with non-radioactive buffer in flow-through chambers. A kinetic analysis of the desaturation curve of ^{45}Ca disappearance from the slices over time can be used to estimate the flow of calcium between three compartments. The fastest of the kinetic phases represents exchange at the extracellular surface, the intermediate phase represents calcium exchange across the plasma membrane, and third or slowest phase probably represents calcium exchange with intracellular pools, including mitochondria.

While our data using this technique are still preliminary, we have observed that 25 μU/ml insulin inhibits the rate of calcium efflux from renal slices (28). This effect of insulin was gradually reduced at the higher concentrations of insulin. The effects of insulin on calcium exchange appear to be localized in the mitochondrial compartment. Further work is needed to determine whether insulin affects specific enzyme systems which are known to play a role in renal calcium transport, and which cellular or subcellular compartments are involved. This would substantially increase our understanding of the regulation of urinary calcium excretion, and of ways in which excessive loss of calcium by this route might be avoided.

Acknowledgments

The author wishes to acknowledge with gratitude the individuals who made a major contribution to this research; G. Block, C.J. Gollaher, M.G. Holl, J. Kerstetter and R.J. Wood.

Literature Cited

1. Kunerth, B.L.; Pittman, M.S. J. Nutr. 1939, 17, 1616.
2. McCance, R.A.; Widdowson, E.M.; Lehmann, H. Biochem J. 1942, 36, 686-91.
3. Margen, S.; Chu, J.Y.; Kaufmann, N.A.; Calloway, D.H. Amer. J. Clin. Nutr. 1974, 27, 584-9.
4. Allen, L.H.; Oddoye, E.A.; Margen, S. Amer. J. Clin. Nutr. 1979a, 32, 741-9.
5. Linkswiler, H.M.; Joyce, C.L.; Anand, C.R. Trans. N.Y. Acad. Sci. 1974, 36, 333-40.
6. Allen, L.H.; Bartlett, R.S.; Block, G.D. J. Nutr. 1979b, 109, 1345-50.
7. Schuette, S.A.; Zemel, M.B.; Linkswiler, H.M. J. Nutr. 1980, 110, 305-15.
8. Pullman, T.N.; Alving, A.S.; Dern, R.J.; Landowne, M. J. Lab. Clin. Med. 1954, 44, 320-32.
9. Moise, T.S.; Smith, A.H. Arch. Pathol. 1947, 4, 530-42.
10. Walser, M.; Mudge, G.H. In "Mineral Metabolism I"; Comar, C.L.; Bronner, F., Eds.; Academic Press: New York, 1960; p. 326.
11. Lemann, J.; Relman, A.S. J. Clin. Invest. 1959, 38, 2215-23.
12. Lemann, J.; Litzow, J.R., Lennon, E.J. J. Clin. Invest. 1967, 46, 1318-28.
13. Witing, S.J.; Draper, H.H. J. Nutr. 1980, 110, 212-22.
14. Block, G.D.; Wood, R.J.; Allen, L.H. Amer. J. Clin. Nutr. 1980, 33, 2128-36.
15. Zemel, M.B.; Schuette, S.A.; Hegsted, M.; Linkswiler, H.M. J. Nutr. 1981a, 111, 545-52.
16. DeFronzo, R.A.; Goldberg, M.; Agus, Z.S. J. Clin. Invest. 1976, 58, 83-90.
17. Heggeness, F.W. J. Nutr. 1959, 142-6.
18. Lindemann, R.D.; Adler, S.; Yiengst, M.J.; Beard, E.S. J. Lab. Clin. Med. 1967, 70, 236-245.
19. Lemann, J., Jr.; Lennon, E.J.; Piering, W.R.; Prien, E.L.; Ricanati, E.S. J. Lab. Clin. Med. 1970, 75, 578-85.
20. Thom, J.W.; Morris, J.E.; Bishop, A.; Blacklock, N.J. Brit. J. Urol. 1978, 50, 459-64.
21. Allen, L.H.; Block, G.D.; Wood, R.J.; Bryce, G.F. Nutr. Res. 1981, 1, 3-11.
22. Holl, M.; Gollaher, C.J.; Allen, L.H. Fed. Proc. (abstract) 1984, 43, 989.
23. Wood, R.J.; Allen, L.H. J. Nutr. 1983, 113, 1561-67.
24. Northrup, T.E.; Krezowski, P.A.; Palumbo, P.J.; Kim, J.K.; Hui, Y.S.F.; Dousa, T.P. Am. J. Physiol. 1979, 236, E649-54.
25. Gollaher, C.J.; Wood, R.J.; Holl, M.; Allen, L.H. J. Nutr. 1984a, 114, 622-626.
26. Roy, D.R.; Seely, J.F. Am. J. Physiol. 1981, 240, F17-24.
27. Zemel, M.B.; Linkswiler, H.M. J. Nutr. 1981b, 111, 315-24.

28. Gollaher, C.J.; Allen, L.H. Fed. Proc. 1984b, 43, 989.
29. Uchikawa,T.; Borle, A.B. Am. J. Physiol. 1978, 234, R34-8.

RECEIVED December 26, 1984

Effect of Dietary Protein, Calcium, and Phosphorus on Calcium Metabolism in Humans

JULIETTE C. HOWE

Energy and Protein Nutrition Laboratory, Beltsville Human Nutrition Research Center, Agricultural Research Service, U.S. Department of Agriculture, Beltsville, MD 20705

Although protein-induced hypercalciuria appears to result from decreased fractional renal tubular reabsorption of calcium, the mechanism by which dietary protein influences this action remains unclear. Utilizing a postprandial response technique, the response of human subjects to meals varying in protein source and/or level, calcium level or phosphorus level was studied. In men, hyperinsulinemia was elicited when cottage cheese, but not beef, was the protein source; hyperinsulinemia did not occur when Ca or P levels were varied. Enhanced calciuria was noted following high protein (cottage cheese) and high beef/high calcium meals in men. The calciuretic response of postmenopausal women was more dramatic than the response of men when high protein as cottage cheese or beef was fed. Serum Ca and P were also more responsive. Serum calcium increased following the ingestion of meals containing 15 g of protein from cottage cheese or beef, or 45 g of protein from cottage cheese. Serum phosphorus levels were lower following high protein meals than after meals containing 0 or 15 g protein. Postmenopausal women may be more susceptible to protein-induced calciuria than younger men.

It has been well established that the ingestion of high dietary protein levels results in hypercalciuria in man, and that hypercalciuria is frequently accompanied by negative calcium balance (1-3). In a summary of data from nutritional surveys in the U.S., Pao (4) showed that dietary protein intake was well above the Recommended Dietary Allowances (RDA) for both men and women regardless of age (5). Although dietary calcium intakes are generally at the RDA for men, women below the age of 50 yr consume only 75% of the RDA (4). Women above the age of 50 yr consume only two-thirds of the RDA for calcium (4). These low consumptions become critical when we consider the reduced ability for calcium absorption demonstrated in both men and women over the age of 60 yr (6).

Although dietary levels of phosphorus are not considered to significantly affect calcium absorption (7), high dietary phosphorus has been shown to reduce urinary calcium levels in human beings (8-12). Studies have in fact demonstrated a reversal of protein-induced calciuria and negative calcium balance when high levels of both calcium and phosphorus were fed with high protein (13-14). The dietary surveys indicate that phosphorus intake is above the RDA for all age groups (4).

Allen et al. (15) have demonstrated a calciuretic effect after consumption of a single meal containing high levels of protein as cottage cheese. These workers have also suggested that insulin may be involved in the calciuretic response to protein ingestion (16). The two studies reported here were designed to corroborate these findings and extend them to other protein sources and levels and to investigate the postprandial response to calcium and phosphorus levels in meals on several parameters associated with calcium metabolism.

Postprandial Response of Men

Methods and Experimental. In the study reported by Beecher et al. (17), nine men aged 23-39 years served as subjects. On the day prior to the initiation of the first experiment, the subjects selected their day's meals from a limited menu. The quantity and choice of foods were then repeated on each subsequent pre-experimental day. The subjects fasted overnight, voided upon rising and then drank 300 ml of water. Upon arrival at the research center (0730-0830 hrs), fasting blood was drawn and urine collected. The subjects then ingested one of the liquid meals and drank 200 ml of water. The subjects also drank 200 ml of water each 1/2 hour during the duration of the experiment to induce diuresis. Blood was drawn at 1/2, 1, 2, and 3 hours post-meal; urine was collected at each 1/2 hour for 3 hours post-meal. Seven to 14 days elapsed between experiments.

Blood samples were centrifuged at 1000 x g for 20 min at 0-4°. Ionized calcium levels were immediately determined in serum and urine samples using a calcium ion-selective electrode (Ionetics, Inc., Costa Mesa, CA); urine volumes were recorded. The remaining serum and urine were aliquoted for various analyses and stored at -40°. Serum insulin was analysed by radioimmunoassay (Amersham Corp., Arlington Heights, IL). Serum levels of total calcium, phosphorus and creatinine as well as urine creatinine were determined by colorimetric procedures using an automated analyzer (Centrifichem, Baker Instruments Corp., Pleasantville, NY). Glomerular filtration rates (GFR) were calculated from serum and urine creatinine data; GFR = urine creatinine/serum creatinine. Urine total calcium levels were analysed spectrometrically by atomic absorption (Perkin Elmer 2380) using a nitrous oxide-acetylene flame and standard conditions of the manufacturer (18). Urine samples and standards were appropriately diluted in 0.25 M KCl to obviate ionization interferences.

Table I shows the composition of the liquid meals consumed by the subjects. Two cottage cheese meals with different levels of protein but the same levels of calcium and phosphorus were

Table I. Liquid Meal Composition-Study With Male Subjects

Protein Source	Protein g/meal	Ca mg/meal	P mg/meal
Cottage cheese	15	300	569
Cottage cheese	45	300	569
Beef	15	166	308
Beef	45	166	308
Beef	45	166	700
Beef	45	466	308
Beef	45	466	700

Note: Constant components are total energy (400 Kcal), lactose (18 g), sucrose (15 g), Na (490 mg), K (965 mg), Mg (88 mg), and Zn (8 mg).

formulated. The calcium and phosphorus levels in these two meals were set at the levels occurring in the high cottage cheese meal.

A meal of cooked lean ground beef provided 15 g of protein and served as the normal or control beef meal. Four additional beef meals were formulated to provide 45 g of protein and 2 levels each of calcium and phosphorus. The lower levels of calcium and phosphorus in the beef meals, that is 166 mg calcium and 308 mg phosphorus, were determined by the natural levels present in those meals containing 45 g of protein from beef. The high levels of calcium and phosphorus in the beef meals were set at one-third of the values shown by other investigators to reverse high-protein induced calciuria ([19]). Other components of the meals, shown on the bottom of Table I, were maintained as constant as possible.

The sequence of meal consumption was determined by random assignment of diets to subjects. Statistical analysis was performed by a General Linear Models Procedure ([20]) using split-plot in time analysis with the following non-orthogonal contrasts:
(a) 15 g vs 45 g of protein from cottage cheese;
(b) 15 g vs 45 g of protein from beef;
(c) 166 mg vs 466 mg calcium in a high protein meal (45 g of protein from beef);
(d) 308 mg vs 700 mg phosphorus in a high protein meal (45 g of protein from beef);
(e) interaction of calcium and phosphorus levels in a high protein meal (45 g protein from beef).

Critical level for all contrasts was set at $P < 0.05$.

Results. Table II shows the postprandial insulin response of young men to the diets. In all cases, insulin rose following meal ingestion, reached maximum at 1/2 hour post-meal, then decreased for the next 2 1/2 hr. The insulin response was significantly greater following the meal with 45 g of protein as cottage cheese than with 15 g of protein as cottage cheese. The insulin response was not different for these two levels of protein when beef was the

Table II. Postprandial Effects of Diet on Serum Insulin Levels
of Men

	Postprandial time (hours)				
Diet	0	1/2	1	2	3
Cottage Cheese					
15g Protein	19[a]	56	29	24	23
45g Protein	16	76	47	30	19
Beef					
15g Protein	18	60	44	26	18
45g Protein	16	57	40	27	19
High Beef					
166mg Calcium	17	58	39	32	22
466mg Calcium	15	56	35	26	21
308mg Phosphorus	15	55	39	25	18
700mg Phosphrous	17	58	35	32	24

[a]All values represent mean in $\mu U/ml$ for 9 subjects; SD (square root of the error mean square) for the ANOVA = 20.44. Responses over all times was significantly different between protein levels when cottage cheese was the protein source ($P < 0.05$).

protein source. In addition, insulin responses were similar when
either the calcium or phosphorus level was varied in a high beef
meal (166 mg Ca vs 466 mg Ca; 308 mg P vs 700 mg P).

Table III gives the quantities of calcium (both total and
ionized) excreted in the urine of men following the ingestion of
meals varying in either protein level or protein source. Regardless
of diet, calcium excretion, both ionized and total, increased
following meal ingestion, reached maximum between 1 and 2 hours
post-meal and then decreased. Feeding 45 g of dietary protein as
cottage cheese did increase urinary total calcium excretion above
those levels obtained following the lower cottage cheese meal. This
response, however, was not statistically significant ($P = 0.06$).
When ionic or free urinary calcium was evaluated, high protein meals
resulted in equal or slightly depressed calciuric response. The
levels of ionic calcium in the urine, as determined by a calcium
ion-selective electrode, suggest that a considerable amount of
urinary calcium was complexed to anions or compounds with anionic
groups.

The urinary calcium response to high levels of protein from
beef showed some reduction in urinary total calcium excretion ($P <
0.06$). Similarly, the excretion of urinary free calcium, was also
depressed when a meal containing high levels of beef was ingested
(Table III).

The effects of varying either the calcium or phosphorus level
in conjunction with a high beef meal on the urinary calcium
excretion of men are shown in Table IV. Urinary calcium excretion
(total and ionized) was significantly elevated ($P < 0.005$) when the
high protein beef meal contained 466 mg rather than 166 mg calcium.
Increasing the phosphorus level from 308 mg to 700 mg in the high
beef meal reduced both total and ionized calcium excretion in the
urine, but the response was not statistically significant. Serum
levels of calcium (ionized and total) and phosphorus were within
normal limits and were unaffected by any of the dietary treatments.

Discussion. The postprandial response of men to meals varying in
protein level was different when the protein source was cottage
cheese rather than beef. When cottage cheese was the protein
source, consumption of a high protein meal did result in a
hyperinsulinemic response. This response was accompanied by a
slight increase in calciuria which confirms the findings of Allen et
al. (16). However, the responses of serum insulin and urinary
calcium excretion to meals containing high levels of beef were quite
different from the responses we observed after the ingestion of high
levels of cottage cheese. In fact, the response of calcium
excretion is directly opposite. One cannot make direct comparisons
on the magnitude of the calciuretic response between the cheese and
beef meals since the protein effects were confounded with
differences in calcium and phosphorus content. However, the level
of several amino acids in these meals have been calculated and show
that the meals containing beef had higher arginine and histidine
levels than the respective meals of cottage cheese. Thus, the
difference in the insulin response to the two protein sources cannot
be simply explained by the levels of these insulinogenic amino acids
in the meal (arginine and histidine). Sulfur amino acids have also
been implicated as a responsible factor in protein-induced

Table III. Postprandial Effects of Protein Level and Source on
Urinary Calcium Excretion of Men

Protein		Postprandial Time (hours)					
Level/Source		1/2	1	1 1/2	2	2 1/2	3
		Total Calcium (mg/30 min)[a]					
15g	Cheese[b]	5.3	5.3	6.0	4.5	4.1	3.7
45g	Cheese	6.0	6.8	6.2	6.6	5.6	3.9
15g	Beef	5.1[c]	5.7	6.7	6.9	6.1	5.3
45g	Beef	5.1	5.3	6.6[c]	6.3	5.5	4.4
		Ionized Calcium (mg/30 min)[d]					
15g	Cheese	2.4	3.7	4.9	4.4	3.4	2.8
45g	Cheese	2.2	3.8	4.1	4.3	3.4	2.6
15g	Beef	2.2[e]	4.0	5.4	5.3	4.4	4.1
45g	Beef	1.9	2.8	4.2	4.2	4.0	3.0

[a]All values represent mean for 9 subjects unless otherwise
specified. SD (square root of error mean square) for the ANOVA =
20.41. Responses over all times between protein levels were nonsignificant for either cottage cheese or beef.
[b]Cottage cheese.
[c]$n = 8$.
[d]All values represent mean for 8 subjects unless otherwise
specified.
[e]$n = 7$.

Table IV.   Postprandial Effects of Ca and P Levels in High Protein
            Diets on Urinary Calcium Excretion of Men

Diet	n	\multicolumn{6}{c}{Postprandial time (hours)}
		1/2	1	1 1/2	2	2 1/2	3
\multicolumn{8}{c}{Total Calcium (mg/30 min)[a]}
166mg Ca	18	4.66	4.61[b]	6.01	5.68	5.12[b]	4.18
466mg Ca	18	5.17	6.03	8.45	7.64	7.78	7.14
308mg P	18	5.32	5.33[b]	7.47	7.14	6.67	6.02
700mg P	18	4.50	5.36	6.99	6.19	6.30[b]	5.30
\multicolumn{8}{c}{Ionized Calcium (mg/30 min)}
166mg Ca	16	1.79	2.57	3.78	3.78	3.22	2.83
466mg Ca	14	2.47	3.28	5.20	5.15	5.27	4.97
308mg P	15	2.06	2.57	4.68	4.83	4.76	4.12
700mg P	15	2.15	3.23	4.20	4.01	3.63	3.54

[a]Values represent combined data for calcium effect and
phosphorus effect since calcium by phosphorus interaction was
determined to be nonsignificant by ANOVA. Responses over all times
were significantly different between calcium levels ($P < 0.005$), but
not between phosphorus levels. SD (square root of the error mean
square) for the ANOVA = 20.41.
    [b]n = 17.

calciuria. In long-term studies, the addition of sulfur amino acids to the diet has been shown to increase calciuria in human subjects, but not to the same extent as that produced by increased protein intake (21). In this study, however, the sulfur amino acid level was similar in the respective meals of beef and cottage cheese.

Increasing the dietary calcium level in the high beef meal resulted in hypercalciuria. This effect was obtained in the absence of an altered insulin response which suggests that factors other than or in addition to serum insulin were involved in the control of urinary calcium excretion.

Long-term consumption of protein has been reported to increase glomerular filtration rate (GFR) and thereby decrease fractional reabsorption of calcium. Data by Allen et al. (15) showed no change in the glomerular filtration rate, but showed a decrease in the percentage of filtered calcium which was reabsorbed by the kidney when a high protein meal was consumed. Under the conditions of this study, feeding high levels of protein did not significantly affect the GFR, which confirms the findings of Allen et al. (15). If high protein intake increases GFR, such an increase cannot be detected within the 3 hr time frame of these postprandial studies.

Postprandial Response of Postmenopausal Women

Experimental. A second study was conducted with nine postmenopausal women age 51-65 yr. The subjects were fed standardized meals for 19 weeks. The mean composition for the 7-day menus of natural foods as % of total calories was 15% protein, 50% carbohydrate, 35% fat with a P/S ratio of 0.7, 10 g/day crude fiber, and less than 300 mg/day cholesterol. In addition, the diets supplied 1289 mg calcium, 1832 mg phosphorus, 2561 mg sodium and 5099 mg potassium daily. The diets met the RDA for all other nutrients. Calorie levels were adjusted to maintain body weight. The experimental meals were fed during the last six weeks of this 19-week period. No more than one liquid meal was consumed by each subject in one week. Fasting and postprandial samples of blood and urine were collected as in the previous study. Diuresis was induced by scheduled consumption of water.

Table V shows the composition of the liquid meals for this study. As in the previous study, two levels each of cottage cheese and beef were fed to the subjects. In addition, 45 g of protein from soy isolate was also fed. The phytic acid content of this meal was 126 mg. A basal diet containing all nutrients except protein was also included in this study. It was necessary to reduce the fat content of the basal diet for palatability; the total energy for this meal only was 250 kcal. In this study, the calcium and phosphorus levels of each meal were equalized across all levels and sources of protein. Levels of other meal components were maintained as constant as possible and are shown at the bottom of Table V.

Meals were consumed by the subjects following a Latin Square Design. Statistical analysis was performed by a General Linear Models Procedure (20) using split-plot in time analysis with the following non-orthogonal contrasts:
(a) 15 g vs 45 g protein from either beef or cottage cheese;
(b) beef vs cottage cheese at either protein level (15 g or 45 g);

Table V. Liquid Meal Composition -- Study with Postmenopausal
Female Subjects

Protein Source	Protein g/meal	Ca mg/meal	P mg/meal
Cottage cheese	15	288	446
Cottage cheese	45	288	446
Beef	15	288	446
Beef	45	288	446
Soy Isolate[a]	45	288	446
No Protein (basal)[b]	0	288	446

Note: Constant components are total energy (452 Kcal), lactose
(18 g), sucrose (15 g), Na (1455 mg), K (580 g), Mg (88 mg), and
Zn (7.3 mg).
[a]Phytic acid content 0.126 g/meal.
[b]Fat content reduced for palatability; total energy was 250 kcal.

(c) interaction of protein source (beef, cottage cheese) and level
     (15 g, 45 g);
(d) beef vs cottage cheese vs soy in a high protein (45 g) meal;
(e) protein vs no protein in a meal.

Critical level for all contrasts was set at $P < 0.05$.

Results. Table VI gives the serum levels of calcium (total and
ionized) and phosphorus. Serum ionized calcium, which ranged from
33% to 36% of total serum calcium, did not respond postprandially to
any of the diets consumed.
     When compared to the basal meal (no protein meal), total serum
calcium levels were significantly elevated following the ingestion
of either cottage cheese meal ($P < 0.05$). Feeding 15 g of protein
from beef also elevated total serum calcium above levels obtained
with the basal diet ($P < 0.05$). Total serum calcium responded
similarly to meals containing 0 or 45 g of protein from either beef
or soy. All high protein meals reduced serum phosphorus to levels
significantly below those obtained following the basal meal ($P <
0.05$). Serum phosphorus increased to a maximum level at 1 hour post
meal, then approached fasting levels by 3 hours post-meal. The
postprandial response of serum phosphorus to the 15 g protein meals
were intermediate to that of 0 or 45 g of protein, but the response
was not significantly different from that following the basal meal.
Statistical evaluation did show however that, regardless of protein
source, serum phosphorus was significantly lowered when 45 g rather
than 15 g of protein was fed.
     Table VII shows the urinary excretion of ionized calcium
following the various liquid meals. Urinary excretion of ionized
calcium increased during the first hour after ingestion of the basal
meal. The excretion rate then decreased and approached fasting
levels by 2 hours post-meal. Although the addition of 15 g of
protein as cottage cheese increased the quantity of ionized calcium

Table VI. Postprandial Effects of Protein Level and Source on
Selected Serum Constituents of Postmenopausal Women

Protein	Postprandial Time (hours)				
Level/Source	0	1/2	1	2	3
Serum Total Calcium (mg/dl)[a]					
15g Beef	10.85[b]	11.07	10.67	10.54	10.38
45g Beef	10.73	10.64	10.50	10.40	10.11
15g Cheese[c]	10.58	10.91	10.67	10.48	10.87
45g Cheese	10.40	10.86	10.73	10.80	10.39
45g Soy	10.76	10.60	10.80	10.29	10.22
0 Basal	10.56[b]	10.49[b]	10.33	10.08	10.54
Serum Ionized Calcium (mg/dl)[d]					
15g Beef	3.60[b]	3.73	3.73	3.78	3.74
45g Beef	3.73[b]	3.71	3.71	3.62	3.75
15g Cheese	3.62	3.71	3.74	3.72	3.75
45g Cheese	3.72	3.68	3.73	3.76	3.67
45g Soy	3.61	3.71	3.63	3.69	3.62
0 Basal	3.62	3.69	3.63	3.62	3.79
Serum Phosphorus (mg/dl)[e]					
15g Beef	3.62[b]	3.58	3.56	3.83	3.76
45g Beef	3.52	3.33	3.27	3.39	3.62
15g Cheese	3.51	3.76	3.59	3.92	3.74
45g Cheese	3.57	3.49	3.34	3.51	3.69
45g Soy	3.62	3.47	3.43	3.63	3.54
0 Basal	3.54[b]	3.64[b]	3.92	3.88	3.67

[a]Responses over all times differed significantly between Basal
and each of the following meals (P < 0.05): cottage cheese (15 g
and 45 g); beef (15 g). SD (square root of the error mean square)
for the ANOVA = 0.475.

[b]Mean for 8 subjects; n = 9 for all other values.

[c]Cottage cheese.

[d]Responses over all times were similar for all diets. SD
(square root of the error mean square) for the ANOVA = 0.2982.

[e]Responses over all times differed significantly (P < 0.05)
between Basal and each of the 45 g protein meals (beef, cottage
cheese, soy); response to meals containing 45 g of protein differed
significantly (P < 0.05) to those containing 15 g of protein. SD
(square root of the error mean square) for the ANOVA = 0.365.

Table VII. Postprandial Effects of Protein Level and Source on
Urinary Ionized Calcium Excretion in Postmenopausal Women

Protein		Postprandial Time (hours)					
Level/Source		1/2	1	1 1/2	2	2 1/2	3
15g	Beef	1.3[a]	3.1	3.3	1.5	1.9	2.0
45g	Beef	1.3	3.9	4.4	3.5	2.6	2.6
15g	Cheese[b]	1.5	2.9	3.7	2.8	1.9	1.8
45g	Cheese	1.7	2.7	5.3	4.6	3.3	3.5
45g	Soy	1.9	3.5	3.6	3.0	2.8	2.2
0	Basal[c]	1.4	2.4	2.1	1.1	1.1	1.0

[a]All values represent means in mg/30 min for 9 subjects.
[b]Cottage cheese.
[c]Responses over all times differed significantly ($P < 0.05$)
between Basal and the high protein cottage cheese and beef meals; SD
(square root of the error mean square) for the ANOVA = 3.182.

excreted, these levels did not differ statistically from those
following the basal meal. The maximum excretion occurred at 1-1/2
hours post-meal. The ingestion of 45 g of protein as cottage cheese
significantly increased the excretion of ionized calcium above that
for the basal meal; maximum excretion was at 1-1/2 hours post meal.
The differences in excretion between 15 g and 45 g of protein as
cottage cheese were not significant.

Significantly ($P < 0.05$) greater quantities of ionized calcium
were excreted in the urine when the high beef rather than the basal
meal was fed. The excretion of ionized calcium following a meal
containing 15 g of beef protein was intermediate to those obtained
after meals containing 0 and 45 g of protein but did not differ
statistically from either one. The excretion of ionized calcium
following the high soy meal did not differ significantly from that
following the basal meal.

Discussion. In this study of postmenopausal women, serum levels of
ionized calcium were unaffected by meal consumption. Levels of
total calcium, however, were elevated following meals containing 15
g of protein from either cottage cheese or beef, or 45 g of protein
from cottage cheese. Serum total calcium was not affected by diet
in the study with men nor in the study reported by Allen et al.
using men and women (16).

Serum phosphorus levels responded more dramatically to the
meals than did serum calcium. Ingestion of a high protein meal
significantly reduced serum phosphorus below those levels obtained
when 0 or 15 g of protein was fed. Ingestion of the basal meal
elevated serum phosphorus. Therefore, the addition of 45 g of
protein to the nutrients in the basal diet negated and reversed the
effect on serum phosphorus. Hypophosphatemia has been shown to

stimulate the synthesis of 1,25-dihydroxycholecalciferol in the kidney, which in turn enhances intestinal absorption of calcium. Since parathyroid hormone (PTH) is not stimulated in cases of hypophosphatemia, calcium reabsorption in the kidney does not occur and calcium diuresis ensues (22). The postprandial reduction of serum phosphorus following a high protein meal may stimulate a similar type of mechanism.

One of the objectives of this study with postmenopausal women was to evaluate postprandial calciuretic response of the subjects to a meal containing all the other dietary constitutents but protein. As seen in Figure 1, ionized calcium excretion in urine rose during the first hour after the meal, then proceeded to return to a steady state by 2 hours post-meal. The addition of 45 g of protein to the meal further stimulated the calciuretic response. The magnitude of the calciuretic response was greatest with cottage cheese, followed by beef, and then soy. The response obtained with soy was not statistically different from that of the basal diet. The calciuretic response with cottage cheese confirm previous findings (15-17). When beef was fed, however, greater calciuretic response occurred when 45 g rather than 15 g of protein was fed. This response is exactly opposite to that obtained with male subjects. The total calcium content of those meals were much lower than that consumed by the postmenopausal female subjects. Therefore, direct comparison of these results cannot be made.

Conclusions

It appears from these studies that the postprandial responses of postmenopausal women may differ from those of younger men. In studies in which calciuria has been demonstrated, the detrimental effect of such a condition on bone has been implied. Osteoporosis is defined as a diminution of skeletal mass in which the bone is normally mineralized but the amount of bone tissue is lower than normal. Unfortunately, it is seldom diagnosed unless bone fracture occurs; 30% of skeletal mass may have been lost by this time (23). Osteoporosis occurs more frequently in females then males and has been estimated as the cause of 90% of all fractures in people over the age of 60 yr (28). Therefore, differences in the postprandial responses of postmenopausal women, a population at high risk for osteoporosis, and men may help to identify contributing factors to the etiology of bone loss.

Serum levels of both total calcium and phosphorus were more responsive to meal ingestion in the postmenopausal women then those previously reported in younger women and/or men (15-17). Serum phosphorus was significantly lowered following high protein meals than when meals containing no protein were fed. In addition, the urinary excretion of ionized calcium was substantially greater following high protein meals of beef and cottage cheese than that resulting from a meal without protein. Determination of such parameters as insulin, PTH, and calcium reabsorption in postmenopausal women may serve to further verify results from other investigations (15-17) or may indicate that other mechanisms are in play for postmenopausal women.

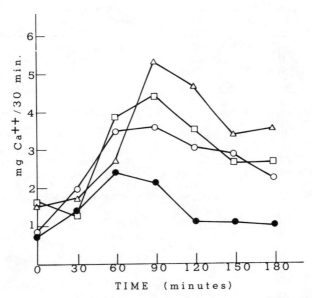

Figure 1.   Urinary excretion of ionized calcium prior to and following a liquid meal.   Key: Δ, 45 g protein, cottage cheese; □, 45 g protein, beef; O, 45 g protein, soy; ●, no protein.

Acknowledgments

The author acknowledges the technical assistance of Robert Staples, Kathleen Fenton, Stephen Majors, Janice Collins, Bernice Brooks, and Rhoda Barnes.

Portions of these studies are from a dissertation to be submitted to the Graduate School, University of Maryland, by Juliette C. Howe, in partial fulfillment of the requirements for the Ph.D. degree in Food, Nutrition, and Institution Administration.

Mention of a trademark or proprietary product does not constitute a guarantee or warranty of the product by the U.S. Department of Agriculture, and does not imply its approval to the exclusion of other products that may also be suitable.

Literature Cited

1. Johnson, N. E.; Alcantara, E. N.; Linkswiler, H. J. Nutr. 1970, 100, 1425-30.
2. Walker, R. M.; Linkswiler, H. J. Nutr. 1972, 102, 1297-02.
3. Anand, C. R.; Linkswiler, H. M. J. Nutr. 1974, 104, 695-0.
4. Pao, E. M.; Mickle, S. J. Food Technol. 1981, 35, 58-79.
5. "Recommended Dietary Allowances," National Academy of Sciences, 1980, 9th ed.
6. Bullamore, J. R.; Gallagher, J. C.; Wilkinson, R.; Nordin, B. E. C. Lancet 1970, 2, 535-7.
7. Allen, L. H. Am. J. Clin. Nutr. 1982, 35, 783-08.
8. Lutwak, L.; Laster, L.; Gitelman, H. J.; Fox, M.; Whedon, G. D. Am. J. Clin. Nutr. 1964, 14, 76-2.
9. Goldsmith, R. S.; Ingbar, S. H. New Engl. J. Med. 1966, 274, 1-14.
10. Spencer, H.; Menczel, J.; Lewin, I.; Samachson, J. J. Nutr. 1965, 86, 125-1.
11. Bell, R. R.; Draper, H. H.; Tzeng, D. Y. M.; Skin, H. K.; Schmidt, G. R. J. Nutr. 1977, 107, 42-0.
12. Spencer, H.; Kramer, L.; Osis, D.; Norris, C. J. Nutr. 1978, 108:447-57.
13. Zemel, M. B.; Linkswiler, H. M. J. Nutr. 1981, 111, 315-4.
14. Linkswiler, H.; Zemel, M. B.; Schuette, S.; Hegsted, M. In "Beltsville Symposia in Agricultural Research 4. Human Nutrition Research"; Beecher, G. R., Ed.; Allenheld, Osmun Publishers: Totowa, 1981; pp. 165-173.
15. Allen, L. H.; Bartlett, R. S.; Block, G. D. J. Nutr. 1979, 109, 1345-0.
16. Allen, L. H.; Block, G. D.; Wood, R. J.; Bryce, G. F. Nutr. Res. 1981, 1, 3-11.
17. Beecher, G. R.; Howe, J. C.; Lakshmanan, F. L.; Bodwell, C. E. Fed. Proc. 1982, 41,771.
18. Perkin-Elmer Corp. "Analytical Methods for Atomic Absorption Spectrophotometry"; Perkin-Elmer Corp: Norwalk, CT, 1976.
19. Linkswiler, H. M.; Zemel, M. B.; Hegsted, M.; Schuette, S. Fed. Proc. 1981, 40, 2429-3.
20. SAS Institute Inc. "SAS User's Guide, 1979 Edition"; SAS Institute Inc.: Cary, NC, 1979; pp. 237-263.

21. Zemel, M. B.; Schuette, S. A.; Hegsted, M.; Linkswiler, H. M. J. Nutr. 1981, 111, 545-2.
22. DeLuca, H. F. Federation Proc. 1974, 33, 2211-9.
23. Stevenson, J. C.; Whitehead, M. I. Br. Med. J. 1982, 285, 585-8.
24. Recker, R. R. Contemporary Nutr. 1983, 8(5).

RECEIVED January 19, 1985

Calcium Metabolism, Bone Mineral Loss, and Soft Tissue Calcification in Rodents

Effect of Phosphorus and Fluoride

B. G. SHAH and B. BELONJE

Banting Research Centre, Health and Welfare, Nutrition Research Division, Ottawa, Ontario, Canada, K1A 0L2

Young and adult male Wistar rats were fed for 6-7 weeks, diets containing 1.3% Ca and 0.6% or 1.2% P. Balance studies were carried out during the last week using ^{45}Ca. From the results it was concluded that the young rats were able to partly overcome the adverse effect of high P intake on Ca metabolism, experienced by adult animals. The effect of fluoride on Ca metabolism was investigated in weanling male rats fed 200 mg/kg fluoride. Fluoride did not have any significant effect on the true absorption or retention of Ca. The turnover of femur Ca was also not affected. Kidney Ca level tended to increase in the fluoride-fed rats. In another experiment, however, 50 mg/kg fluoride decreased kidney calcification but a simultaneous increase in the magnesium level from 450 to 2700 mg/kg diet was required to prevent nephrocalcinosis observed in male and female rats, fed either NRC or AIN mineral mixtures. The interaction of dietary fluoride with low and high levels of Ca but a constant level of P was investigated in another 2 yr experiment on mice. The results indicated that: 1, The low Ca diet resulted in lower total femur Ca than the comparable values for the high Ca group, whether fluoride was fed or not; 2, there was no significant effect of fluoride on the breaking load of femur shaft at either level of calcium.

The problem of bone mineral loss, which accompanies aging of both men and women, has recently received much attention of the scientific community because the proportion of the elderly in the population is increasing especially in Western Europe and North America (1,2). Among the various factors which affect the integrity of the skeleton are habitual intakes of mineral nutrients such as calcium, phosphorus and fluoride. Here we report the results of our investigations on the interactions between these minerals in relation to bone and soft tissues.

0097-6156/85/0275-0141$06.00/0

Calcium and Phosphorus

Although a clear relationship between calcium intake and bone
status has not been definitely established, an association between
chronic low calcium intake and osteoporosis in women was reported.
Calcium supplement has also been shown to improve their skeletal
status (3). The ratio of P:Ca was found to have an adverse effect
on bone only when habitual intake of calcium was below 50% of the
recommended dietary allowance (3). This effect is caused by secondary hyperparathyroidism in experimental animals but in man,
although increased parathyroid activity has been demonstrated, the
adverse effect on bone has not been shown (1). In man (4) and in
the rat (5), a high phosphorus intake did not have any effect on
calcium absorption, when calcium gluconate and sodium glycerophosphate were used as calcium and phosphorus supplements in the
human experiments and the level of Ca in the rat diet was 0.6%, a
little more than recommended by the American Institute of Nutrition
(6). When inorganic sources of Ca and P (calcium carbonate,
calcium or sodium phosphate) were used, however, high P resulted in
a decreased calcium balance in adult women in spite of high Ca
intake (7). In the rat, the effect of high P on Ca absorption in
the presence of a high level of dietary Ca was not known. Moreover, it was necessary to determine whether age had any effect on
the interaction between Ca and P at high levels. The results of
such an investigation in young rats were reported by us (8). We
observed that high P in the diet did not affect the bone turnover
but reduced the true absorption of Ca by 16%. Here we report the
results of a similar experiment on adult rats to determine the
effect of age.
        Sixty 6-month old male Wistar rats weighing 525 g were
injected intraperitoneally with $^{45}CaCl_2$ and fed for 2 weeks a
closed formula diet containing 2.1% Ca and 1.1% P. Six randomly
selected rats were then killed (0 time) and blood and femur collected. The remaining rats were randomly divided into two equal
groups and one group was fed for 6 weeks a diet containing 1.3% Ca
and 0.6% P and the other a diet containing the same level of Ca but
double the P. After 2 and 4 weeks, 8-9 rats from each group were
killed and a metabolic study conducted on the remaining 10 rats
during the sixth week of feeding. The details of diets and methods
were the same as described before (8). The results obtained with
the adult rats are compared to those in the young rats in Tables I
and II.
        It is evident from the data in Table I that in the young rats
high dietary phosphorus caused a 54% decrease in urinary Ca and a
slight (14%) reduction in Ca retention, inspite of lower endogenous
fecal excretion. The true absorption of calcium was reduced by 15%
only. In the adult rats, however, the higher intake of P resulted
in a slight increase in fecal Ca but no change in urinary Ca or
endogenous fecal Ca and a small loss of body Ca. The decrease in
true absorption (47%) in adult rats was three times that in the
young rats indicating that the former could not maintain Ca balance
as the young rats did (8). The effect of high P on urinary Ca

Table I. Calcium absorption in young and adult rats

Parameter		Young rats		Adult rats	
Age (weeks)		8		33	
Dietary calcium (%)		1.3	1.3	1.3	1.3
phosphorus (%)		0.6	1.2	0.6	1.2
		Calcium (mg/day)			
Intake		234 ± 10^a	225 ± 9	196 ± 4	195 ± 7
Excretion:	Feces	177 ± 9	176 ± 9	192 ± 6	211 ± 5^c
	Urine	1.3 ± 0.2	0.6 ± 0.1^b	0.5 ± 0.1	0.5 ± 0.1
Balance		56 ± 2	48 ± 2^c	5 ± 2	-10 ± 2^b
Endogenous fecal		20 ± 2	13 ± 1^b	29 ± 2	27 ± 3
		percent			
True absorption		33 ± 2	28 ± 1^c	17 ± 2	9 ± 1^b

[a]Mean $\pm$ standard error (N=9/group).

[b,c]Significantly different from the low phosphorus diet: b (P< 0.01);
c (P < 0.05).

Table II. ^{45}Calcium in femur ends of young and adult rats

Parameter	Young rats		Adult rats	
Age (weeks)	8		33	
Dietary calcium (%)	1.3	1.3	1.3	1.3
phosphorus (%)	0.6	1.2	0.6	1.2
Time after injection (days)	^{45}Ca (% dose)			
0	3.29±0.09[a]	-	2.09±0.10	-
14	-	-	1.89±0.05	1.55±0.07[b]
17	2.90±0.06	2.89±0.08	-	-
24	2.70±0.06	2.72±0.07	-	-
28	-	-	1.53±0.02	1.55±0.04[b]
33	2.85±0.08	2.70±0.06	-	-
42	-	-	1.44±0.06	1.20±0.05[b]
Half life (days)	168		73	60[b]

[a]Mean ± standard error (N=9/group).

[b]Significantly different from the low phosphorus diet ($P < 0.01$).

observed by us in young rats was similar to that in adult rats fed
a diet containing 0.6% (5) but urinary Ca in adult rats fed 1.3% Ca
did not change. Moreover, Draper et al. (5) did not observe any
effect of high P on net absorption of Ca by adult rats but our
calculation of true Ca absorption revealed an appreciable decrease.
It is likely that high dietary P does not have the same effect when
the diet does not contain high Ca also. When the Ca intake of men
was low or marginal, high P did not have any effect on Ca balance
(9) but women consuming high Ca and P experienced much lower Ca
retention than when the dietary P was lower (7).

There was no significant effect of high dietary P on the half
life of Ca in femur ends of young rats (Table II) but the turnover
was faster in adult rats fed high P. This observation is similar
to that reported earlier and was probably due to secondary hyperparathyroidism (5). Thus, our results indicated that age had an
adverse effect on the ability to adapt to high dietary P in the
presence of high Ca. In the human, age has been reported to blunt
the adaptive response to a low Ca intake (10) but a similar effect
of aging on adaptation to high dietary Ca and P from inorganic
sources (7) remains to be confirmed by current methodology
involving the use of isotopes.

Fluoride and Calcium

Based on limited epidemiologic evidence, fluoride supplements, with
or without calcium, estrogen and vitamin D, are used by clinicians
for the treatment of osteoporosis. However, knowledge of the
effects of fluoride on calcium and phosphorus metabolism in normal
animals is limited although Spencer et al. (32) reported that
ingestion of fluoride by three osteoporotic men did not affect
calcium absorption but caused a decrease in urinary excretion.
Moreover, there is a need to determine the long-term effects of
fluoride treatment on bone strength and on soft tissues (2).

In a 6-week experiment on young rats we investigated the
effects of fluoride on calcium-phosphorus metabolism, turnover of
bone calcium and on bone strength (11). The results are summarized
in Table III-V. Adding 200 mg F/kg (as NaF) to a diet containing
1.3% Ca and 0.6% P, did not have any effect on absorption or
excretion of Ca (Table III). There was, however, a slightly
increased P retention in the fluoride-fed rats (results not given
here). The lack of any effect of fluoride on Ca absorption was
later confirmed by in vitro transport and absorption of ^{45}Ca in
everted intestines from rats fed a diet containing 1% Ca (12) and
agrees with the observation in man (32).

The results in Table IV indicate that although the fluoride
content of bone increased to 200 times, there was no change in the
content or turnover of bone calcium. Dietary fluoride, however,
had a slight adverse effect on the strength of the femur shaft as
measured by the breaking load. A similar adverse effect on the
strength of rat bone was also observed by Guggenheim et al. (13) at
a high (150 mg/L) fluoride level in the drinking water but at a 50
mg/L fluoride level there was no effect. In human patients,
however, the effect of therapeutic doses of fluoride on the

Table III. Effect of fluoride on calcium absorption[a]

Parameter		Dietary fluoride (mg/kg)	
		0.2	200
		Calcium (mg/day)	
Intake		208+8[b]	208+8
Excretion:	Feces	159+5	160+7
	Urine	0.8+0.1	0.8+0.1
Balance		48+4	48+3
Endogenous fecal		8.9+0.5	8.7+0.6
		(percent)	
True absorption		27+2	27+1

[a]The diet contained 1.3% calcium and 0.6% phosphorus. Weanling rats fed for 6 weeks. Phosphorus retention increased 15% due to fluoride feeding.

[b]Mean + standard error (N=12/group).

Table IV. Effect of fluoride on bone composition and strength

Parameter		Dietary fluoride (mg/kg)	
		0.2	200
Calcium (mg/g dry weight)		235+1[a]	233+1
Fluoride (μg/g dry fat-free)		33+3	6650+440
^{45}Ca in femur (%dose):	ends	4.2+0.2[b]	4.3+0.1
	shafts	2.7+0.1	2.6+0.1
Breaking load of shafts (pounds)		94+5	78+5[c]

[a]Mean + standard error (N=12/group).

[b]Thirteen days after I.P. injection of ^{45}Ca.

[c]Significantly different from lower fluoride diet ($P < 0.05$).

Table V. Effect of fluoride on soft tissue levels

Parameter	Dietary fluoride[a] (mg/kg)	
	0.2	200
	Fluoride	
Plasma (μg/ml)	0.23 ± 0.03[b]	0.83 ± 0.13[c]
Liver (μg/g fresh weight)	0.08 ± 0.01	0.26 ± 0.02[c]
Kidney (μg/g fresh weight)	0.12 ± 0.04	0.96 ± 0.10[c]

[a]No effect on phosphorus, but calcium appeared to increase in kidney from 7 to 10 mg/100 g fresh weight.

[b]Mean $\pm$ standard error (N=12/group).

[c]Significantly different from lower fluoride diet ($P < 0.005$).

strength of bones as measured by the incidence of fractures is not consistent (2,3,14).

As can be seen from the results in Table V, fluoride levels in plasma, liver and kidney increased 3 to 8 times but there was no significant effect on the calcium or phosphorus content, although the kidney Ca level in fluoride treated rats was 40% higher than in the controls. Whereas the normal exposure to fluoride from air, food and water did not cause any increase in soft tissue levels, more than ten times the normal levels in soft tissues, including liver and kidney, were found in human fatalities due to fluoride poisoning (15).

In order to determine the effect of long-term fluoride feeding, in conjunction with low or high levels of dietary Ca and a fixed level of P, 200 female 6-week old mice of DBA Jax 2 strain, obtained from Bar Harbor, Maine, were maintained on a closed formula diet for one week and then divided randomly into two equal groups. One group was fed ad lib, a diet containing g/100g: casein, 20; cornstarch, 25; corn oil, 5; alphacel (non-nutritive fibre), 2; vitamin mixture (8), 1; mineral mixture excluding Ca and P (8), 0.65; Ca (H_2PO_4)$_2$ · H_2O, 1.06; $CaCO_3$, 0.06; sucrose, 45.23. This diet contained 0.2% and 0.4% added Ca and P respectively. The other group received the same diet except 1.56 g $CaCO_3$ was added to provide 0.8% Ca and sucrose adjusted accordingly. The mice were housed in individual plastic cages, had free access to tap water containing 1 mg/L fluoride and were weighed once a week. The protocol for the experiment is given in Table VI. At each time from 214 days, randomly selected mice were killed by an overdose of sodium pentobarbital and femur, tibia, heart and kidneys were collected for the determination of ^{45}Ca, Ca, P, F and the breaking load of the femur shaft by methods described earlier (8,11). The results are summarized in Tables VII-XI. Although body weights are not given here, it is pertinent to note that except for the decrease in the body weight with aging beyond 466 days in mice fed

Table VI. Protocol of long-term mouse[a] experiment

| | Diet number | |
	10	20
Dietary calcium (%)	0.2	0.8
phosphorus (%)	0.4	0.4
Number of mice	100	100

AGE (days)	
42	Start feeding
179	Inject ^{45}Ca
214	Kill 6 from each group
376	Kill 4 from each group
466	Kill 6 from each group
	Divide the remaining mice into two subgroups. Feed 50 μg/g fluoride to one subgroup (#11 and #21) from each diet group
527	Kill 3 from each group
590	--------"--------
653	--------"--------
712	Kill the remaining mice

[a]DBA Jax 2 female mice obtained from Bar Harbor, Maine.

both the levels of Ca, no other differences were significant. Thus
the low dietary Ca did not have any adverse effect on growth, nor
did 50 mg F/kg diet have any toxic effect on aging mice.
    The loss of ^{45}Ca from the femur was found to be exponential

$$y = Ae^{Bt}$$

(where y=% dose of ^{45}Ca, A=intercept, B=slope and t=age in days;
half-life=-0.693/B) in all groups (Table VII) and calculation of
the half-life revealed that there was no significant difference
between the turnover of bone Ca of mice fed the low and high Ca
diets up to 15 months of age. With further aging, however, the
half-life of Ca did not change in the low Ca fed mice, but the
turnover decreased significantly in the mice fed the high calcium
diet. Feeding fluoride (50 mg/kg diet) tended to slow the turnover
of bone mineral at both levels of dietary Ca but the differences
were not significant.
    The total Ca in the femur (Table VIII) was not significantly
affected by the two levels of dietary Ca (214 to 466 days) indicating that the adaptive response could compensate for the difference
in Ca intake upto 15 months of age but later on (527 to 712) the
difference became significant (P < 0.008) indicating a decline in
the adaptive capacity with aging (10). Krishna Rao et al. (16)
also observed a decrease in femur Ca of 13 or 25 month old female

Table VII.   [45]Calcium in the femurs of mice

Age (days)	Diet number[a]			
	10	11	20	21
	[45]Ca (% dose)			
214	1.47±0.04[b]		1.24±0.04	
376	0.83±0.04		0.77±0.03	
466	0.63±0.04		0.51±0.03	
527	0.58±0.07	0.52±0.07	0.50±0.06	0.52±0.09
590	0.51±0.05	0.42±0.02	0.44±0.03	0.48±0.07
633	0.33±0.02	0.34±0.02	0.32±0.01	0.36±0.04
712	0.40(1)[c]	0.32±0.05(2)	0.32±0.03	0.36±0.02(2)
	Half life (days)			
214-466	205		198[d]	
527-712	214	237	277[d]	322

[a]For composition of diets see Table VI.   [b]Mean ± standard error.
[c]Number of mice.   [d]Significantly different ($P < 0.05$) from each other.

mice fed 0.4% Ca as compared to those fed 1.2% Ca, the Ca:P ratio
being 2:1 at both levels.

Feeding fluoride did not have a significant effect on femur
calcium at either level of dietary Ca, although the fluoride
content of tibia increased at both levels (Table IX). The accumulation of fluoride in the bone of low Ca fed mice was higher
($P < 0.05$) than that in the mice fed the high Ca diet. A similar
observation was reported in rats (13) and in monkeys (17). The
concentration of fluoride in the tibia of treated mice at the end
of the experiment (not shown in Table IX) was about 2500 µg/g dry
weight in the low Ca group and about 1300 µg/g dry weight in the
high Ca groups. These levels were less than half that (about 5000
µg/g) at which the strength of rat bone was reduced (13). In man
also it is estimated that a physiological hazard to the individual
may not arise at levels of fluoride upto 5000 µg/g skeletal
tissue (18). The breaking load of femur shafts (Table IX) was also
not affected by fluoride feeding. The only significant adverse
effect on bone strength was due to aging of the mice fed the low Ca
diet. The apparent decrease in the strength of femur shafts of
mice fed the high Ca diet was not significant. It can be concluded
from these observations that the high Ca intake during aging,
rather than the intake during growth and early adulthood was important for sufficiently preserving skeletal integrity. This supports

Table VIII. Calcium in the femurs of mice

Age (days)	Diet number[a]			
	10	11	20	21
	Calcium (mg)			
214	11.4±0.2[b]		12.0±0.3	
376	10.9±0.4		12.6±0.5	
466	11.2±0.2		10.4±0.5	
527	9.7±1.1	9.0±1.8	10.5±0.6	10.2±0.8
590	8.7±0.4	8.9±0.6	10.5±0.9	10.4±0.2
653	7.4±0.1	7.8±0.4	9.4±0.4	9.2±0.7
712	8.8(1)[c]	8.3±1.1(2)	9.4±1.0	9.7±1.1(2)
	Slope[d] (mg/day)			
214-266	-0.0011		-0.0057	
527-712	-0.0107	-0.0059	-0.0072	-0.0045

[a]For composition of diets see Table VI.   [b]Mean ± standard error.
[c]Number of mice.   [d]None of the differences were significant.

the observation that even 70 year old osteoporotic females can
benefit from supplementary Ca (19). The lack of any beneficial
effect of long term fluoride ingestion on bone strength does not
support the treatment of senile osteoporotics with fluoride
supplements (3).

Whereas much attention has been paid to the effects of dietary
fluoride on the skeleton of domestic and experimental animals and
man (18), knowledge about the accumulation of fluoride in soft
tissues such as heart, liver or kidney is scarce. The results in
Table X indicate that the fluoride content of the kidneys of mice
increased with age when the dietary Ca was low and fluoride feeding
resulted in much higher levels at 653 days of age irrespective of
the dietary Ca content. It is likely that such high fluoride
levels had an adverse effect on kidney function. In cases of
chronic fluoride intoxication, kidney function was reported to be
impaired in the majority of cases (18). It should be noted that
fluoride accumulated in the kidneys of mice even though the
capacity of bone to store fluoride (5000 µg/g dry weight) without
skeletal damage was not reached.

The kidney Ca level generally increased with age in all groups
and fluoride feeding did not seem to have any consistent effect on
the accumulation of Ca. In young rats, however, we observed an
apparent increase in kidney Ca when 200 mg/kg fluoride was fed for
only 6 weeks (Table V). When the fluoride level in the diet was 50
mg/kg, however, nephrocalcinosis in young rats was found to
decrease (20).

Table IX. Fluoride content[a] and strength[b] of mouse bone

Age (days)	Diet number[c]			
	10	11	20	21
	Fluoride (µg)			
214	10 ± 1[d]		8 ± 1	
376	14 ± 1		11 ± 2	
466	14 ± 0		$11\pm1(5)$[e]	
527	13 ± 1	63 ± 7	12 ± 1	37 ± 4
590	19 ± 3	102 ± 15	15 ± 1	59 ± 7
653	16 ± 1	125 ± 12	16 ± 2	93 ± 5
712	16(1)	$150\pm33(2)$	13 ± 1	$80\pm16(2)$
	Breaking load (pounds)			
214-466	7.73 ± 0.74[f]		7.42 ± 0.69	
527-712	3.70 ± 0.92[f]	4.56 ± 0.93	5.28 ± 0.59	4.24 ± 0.42

[a]Of tibia   [b]Of femur shafts
[c]For composition of diets see Table VI.   [d]Mean ± standard error.
[e]Number of mice.   [f]Significantly different from each other (P<0.01).

Table X. Calcium and fluoride in kidney (µg/g dry weight pooled sample)

Age (days)	Diet number[a]							
	10		11		20		21	
	Ca	F	Ca	F	Ca	F	Ca	F
214	480	1			440	1		
376	480	2			470	1		
527	890	2	650	7	670	2	650	4
653	970	6	890	21	560	1	1250	31

[a]For composition of diets see Table VI.

The Ca content of the heart (Table XI) increased with aging in the low Ca fed mice but this was not so for those fed the high Ca diet. However, atrial and ventricular Ca levels were found to increase in female DBA mice (21) fed a Purina stock diet (1% Ca), although the Ca concentration at 12-16 months of age was only 400 mg/kg fat free dry weight. Even with such a low Ca level, Clower et al. observed an increase in fibrotic and necrotic lesions. Our results in both low and high Ca fed mice were much higher even when fluoride was added to the diet. Values from 3000 to 5000 mg/kg dry weight of heart from 15-month old female rats were reported by Dalderup et al. (22) and they also observed a large increase with age. There is an indication that dietary fluoride decreased the calcium level in the heart of the low Ca fed mice but perhaps had the opposite effect in the mice fed the high Ca diet. On the contrary, high fluoride in drinking water, was reported to be associated with a decreased incidence of osteoporosis in women and of aortic calcification in men (23). Subsequent investigation, however, showed a positive correlation between osteoporosis and aortic calcification irrespective of sex (24).

The fluoride content of the heart appeared to increase in old mice fed the low Ca diet and fluoride feeding caused an increase in the level of the hearts of old mice fed high or low Ca diet. When male or female mice were supplied drinking water containing 10 mg/L fluoride, the levels in the soft tissues including heart and kidney were, however, undetectable at 24 months of age (25). This could be due to limited sensitivity of the method used by these investigators.

Table XI. Calcium and fluoride in heart (μg/g dry weight pooled sample)

Age (days)	Diet number[a]							
	10		11		20		21	
	Ca	F	Ca	F	Ca	F	Ca	F
214	7610	17			6420	10		
376	2940	13			3560	23		
527	11000	8	2300	12	1230	5	5760	47
653	19800	57	13200	125	3920	15	11200	87

[a]For composition of diets see Table VI.

In summary, the results of this life-long experiment on female mice indicated that low or high Ca intake early in life did not have any significant effect on the skeletal Ca in adulthood, but the continued high Ca intake reduced the turnover rate in old age. Consequently the strengh of the bone was better preserved than that in the mice fed the low Ca diet. This supports the use of Ca supplements even in old patients suffering from osteoporosis, provided their vitamin D status is adequate. Dietary fluoride

administration at 50 mg/kg diet from 15 months of age did not
impart any benefit to the skeletal status in old age. Thus the use
of fluoride supplements to prevent or reduce bone mineral loss is
not supported. The accumulation of Ca and fluoride in soft
tissues, such as kidneys and heart of old mice especially when the
dietary Ca was low and its effect on organ function remain to be
investigated.

Nephrocalcinosis, Magnesium, Fluoride

At the end of a 16-week feeding test of rapeseed protein concen-
trates (RPC) we observed (26) that female Sprague-Dawley rats fed
casein or RPC, had high levels of Ca and P in the kidney, but not
the animals which received Master's Lab Chow, a stock diet (Table
XII). It was intriguing to find that the Ca content of some
kidneys was more than 100 times the normal. It was evident that
the nephrocalcinosis in the female rats was associated with
purified diets. A review of the literature (20) revealed that
among minerals, dietary Ca, P, Mg, Cl and F were involved in
nephrocalcinosis. After investigating the effects of adding Mg, Cl
or F to purified diets in different combinations (20, 27) it was
conclusively shown (Table XIII) that nephrocalcinosis could be prevented in male and female rats by supplementing the salt mixtures
recommended either by the National Research Council or American
Institute of Nutrition, with the three elements (27). The addition
of chloride, however, was not essential, at least in the case of
females. Although dietary fluoride (50 mg/kg) alone reduced the
kidney Ca levels in weanling female rats (20) fed for six weeks, a
similar effect was not seen in the long-term experiment on female
mice (Table X). The difference could be attributed to the age or
species of the animals.
     Since the source of protein appeared to affect the maximum
kidney Ca level reached (Table XII) we decided to determine the
effect of feeding, for 6 weeks, soy assay protein or spray-dried
egg white on kidney calcification in rats. The results are summarized in Table XIV. To our surprise kidneys of male and female
rats fed egg white powder showed a normal Ca content. It is likely
that the increased urinary sulphate excretion by rats fed egg white
(28) prevented the precipitation of calcium phosphate in the

Table XII. Kidney calcium in rats fed casein and rapeseed
     protein concentrates for 16 weeks (μg/g dry weight)

Protein	Male	Female
Casein[a]	0.46 (0.31-0.83)[b]	20.76 (14.83-54.35)
Tower RPC[a]	0.32 (0.30-0.40)	0.91 (0.48- 3.26)
Span RPC[a]	0.43 (0.37-0.61)	22.13 (0.43-29.44)
Stock diet	0.36 (0.27-0.43)	0.46 (0.29- 0.57)

[a]With NRC-72 salt mixture.   [b]Median (Range), N=5.

Table XIII. Kidney[a] calcium in rats fed NRC-72 and AIN-76 salt
mixtures with fluoride and extra chloride and
magnesium for 6 weeks (mg/g dry weight)

Diet	Male	Female
NRC-72	0.55 (0.28 - 9.84)[b]	16.97 (2.57 - 45.73)
NRC-72+Cl+Mg+F[c]	0.36 (0.26 - 0.56)	0.25 (0.24 - 0.28)
AIN-76	0.49 (0.26 - 3.44)	9.71 (0.48 - 31.07)
AIN-76+Cl+Mg+F[c]	0.33 (0.27 - 0.51)	0.23 (0.21 - 0.25)
Stock diet[d]	0.26 (0.25 - 0.28)	0.27 (0.23 - 0.58)

[a]Calcium levels in heart and liver were not affected.

[b]Median (Range), N=10.   [c]Total (mg/kg): Cl 6000; Mg 2700; F 50.

[d]Total (mg/kg): Cl 6000; Mg 2700; F 40.

Table XIV. Kidney calcium in rats fed different sources of protein
for 6 weeks (mg/g dry weight)[a]

Sources	Male	Female
Casein	0.49 (0.28 - 3.44)[b]	9.71 (0.48 - 31.07)
Soy	0.30 (0.28 - 1.50)	21.83 (0.27 - 44.95)
Egg white	0.27 (0.23 - 0.31)	0.26 (0.20 - 0.30)
Stock diet	0.26 (0.25 - 0.28)	0.27 (0.23 - 0.59)

[a]Diet contained 20% protein and AIN-76 salt mixture.

[b]Median (Range), N=10.

kidney. Further investigation of the causes and prevention of this
nephrocalcinosis and its effects on kidney function are required,
since it may affect the interpretation of the results of animal
studies designed to evaluate the safety of food constituents and
additives (29). It is also worth noting that in man, distal renal
tubular acidosis usually produces calcium phosphate stones or
nephrocalcinosis (30). Moreover, calcium-stones in the upper
urinary tract occur in individuals of affluent societies and the
incidence in the U.K. was found to parallel the consumption of
animal protein from meat, fish and poultry (31).

Literature Cited

1.   Chinn, H.I. "Effects of Dietary Factors on Skeletal Integrity
      in Adults: Calcium, Phosphorus, Vitamin D, and Protein"; Life
      Sciences Research Office, Federation of American Societies for
      Experimental Biology: Bethesda, Maryland, 1981; pp. 1-75.

2. Anderson, S.A., Ed. "Effects of Certain Vitamins and Minerals on Calcium and Phosphorus Homeostasis"; Life Sciences Research Office, Federation of American Societies for Experimental Biology: Bethesda, Maryland, 1982; pp. 1-93.

3. Albanese, A.A. Biblthca Nutr. Dieta 1983, 33, 80-99.

4. Spencer, H.; Kramer, L.; Osis, D.; Norris, C. J. Nutr. 1978, 108, 447-57.

5. Draper, H.H.; Sie, T.; Bergan, J.G. J. Nutr. 1972, 102, 1133-42.

6. American Institute of Nutrition Ad Hoc Committee on Standards for Nutritional Studies. J. Nutr. 1977, 107, 1340-48.

7. Leichsenring, J.M.; Norris, L.M.; Lamison, S.A. J. Nutr. 1951, 45, 407-18.

8. Shah, B.G.; Méranger, J.C. Can. J. Physiol. Pharmacol. 1970, 48, 675-80.

9. Malm, O.J. "Calcium Requirement and Adaptation in Adult Men"; Oslo University Press: Oslo, Norway, 1958.

10. Bowman, B.B.; Rosenberg, I.H. Human Nutr.: Clin. Nutr., 1983, 37C, 75-89.

11. Deshmukh, D.S.; Méranger, J.C.; Shah, B.G. Can. J. Physiol. Pharmacol. 1970, 48, 503-9.

12. Chang, Y.O.; Pan, M.; Varnell, T. Nutr. Rep. Internat. 1977, 16, 539-547.

13. Guggenheim, K.; Simkin, A.; Wolinsky, I. Calcif. Tiss. Res. 1976, 22, 9-17.

14. Bikle, D.D. Annals Inter. Med. 1983, 98, 1013-1015.

15. "Fluorides", National Academy of Sciences, 1971; pp. 172-6.

16. Krishna Rao, G.V.G.; Ts'ao, K.; Draper, H.H. J. Gerontol. 1972, 27, 183-7.

17. Sriranga Reddy, G.; Srikantia, S.G. Metabol. 1971, 20, 642-9.

18. McClure, F.J. "Water Fluoridation: The Search and the Victory"; U.S. Department of Health, Education and Welfare: Bethesda, Maryland, 1970.

19. Lee, C.J.; Lawler, G.S.; Johnson, G.H. Am. J. Clin. Nutr. 1981, 34, 819-823.

20. Shah, B.G.; Belonje, B.; Nera, E.A. Nutr. Rep. Internat. 1980, 22, 957-963.

21. Clower, B.R.; Ball, C.R.; Williams, W.L. J. Atheroscler. Res. 1968, 8, 637-642.

22. Dalderup, L.M.; Keller, G.H.M.; Stroo, M.M. Gerontologia 1967, 13, 86-94.

23. Bernstein, D.S.; Sadowsky, D.M.; Hegsted, D.M.; Guri, C.D.; Stare, F.J. J. Am. Med. Assoc. 1966, 198, 499-504.

24. Boukhris, R.; Becker, K.L. JAMA 1972, 219, 1307-1311.

25. Schroeder, H.A.; Mitchener, M.; Balassa, J.J.; Kanisawa, M.; Nason, A.P. J. Nutr. 1968, 95, 95-101.

26. Shah, B.G.; Benns, G.; Nera, E.A.; Verdier, P.C.; Beare-Rogers, J.L.; Jones, J.D.; Ohlson, R.; Anjou, K. Qual. Plant. Plant. Foods Hum. Nutr. 1981, 30, 234-243.

27. Shah, B.G.; Belonje, B. Nutr. Res. 1983, 3, 749-760.

28. Calvo, M.S.; Bell, R.R.; Forbes, R.M. J. Nutr. 1982, 112, 1401-1413.

29. Roe, F.J.C. J. Hum. Nutr. 1979, 33, 405-415.

30. Coe, F.L.; Favus, M.J. In "Advances in Internal Medicine";
     Stollerman, G.H., Ed; Year Book Medical Publishers: Chicago,
     1980, Vol.26, pp.373-392.
31. Robertson, W.G.; Peacock, M.; Hodgkinson, A. J. Chron. Dis.
     1979, 32, 469-476.
32. Spencer, H.; Lewin, I.; Osis, D.; Samachson, J. Am. J. Med.
     1970, 49, 814-822.

RECEIVED January 7, 1985

Effect of Certain Minerals on the Bioavailability of Calcium in Adult Males

HERTA SPENCER and LOIS KRAMER

Metabolic Section, Veterans Administration Hospital, Hines, IL 60141

Studies were carried out in adult males to determine the effect of phosphorus, magnesium, fluoride and zinc on the availability of calcium for absorption. Increasing the phosphorus intake from 800 to 2000 mg per day did not affect the intestinal absorption of calcium, determined with tracer doses of ^{47}Ca, and did not change the calcium balance. Increasing the magnesium intake from 230 mg to 850 mg per day by adding magnesium oxide to a constant diet also did not affect the absorption of calcium nor the calcium balance. Adding 45 mg fluoride as sodium fluoride to a dietary intake of 4.2 mg fluoride decreased the urinary calcium but did not affect the absorption of calcium. A high zinc intake of 140 mg per day, given as the sulfate, added to a dietary zinc intake of 15 mg per day decreased the absorption of calcium during a low calcium intake of 220 mg per day, but not during a normal calcium intake of 800 mg per day. These studies have shown that phosphorus, magnesium and fluoride in the doses used had no effect on the bioavailability of calcium, while large doses of zinc given during a low calcium intake inhibited the intestinal absorption of calcium. The effect of protein and of certain drugs on calcium metabolism is also described.

The recommended dietary allowance, RDA, for calcium in the United States is 800 mg per day (1). However, this amount of calcium may not be adequate throughout adult life for maintaining the normal skeletal structure and for preventing bone loss with aging, particularly in females. Recent studies indicate that this amount of calcium may not be adequate for maintaining the optimal bone mass which reaches its peak at the age of 30 to 35 years (2) and decreases thereafter. Even if the calcium intake is adequate, one has to consider to which extent this amount of calcium is utilized. There may be a decrease in the intestinal absorption of calcium (3) or the utilization of calcium may be interfered with by other substances, which may be dietary factors, or by the use of certain drugs (4).

The present communication is concerned with calcium balance studies carried out during the intake of a variety of dietary factors including minerals and protein as well as drugs. The studies were performed in adult males under strictly controlled conditions in the Metabolic Research Unit. The diet was kept constant throughout the relatively long term studies and complete collections of urine and stool were obtained. The composition of this diet was previously described (5). The diet and the excretions in urine and stool were analyzed for calcium throughout the studies. These data formed the basis of calcium balances which were determined for several weeks. Calcium was analyzed by atomic absorption spectroscopy (6).

Dietary Factors Which May Affect the Absorption and Utilization of Calcium

Phosphorus. The main sources of dietary phosphorus are dairy products and meat. The usual dietary phosphorus intake is approximately 1 gm per day. It has been postulated by several investigators that a high phosphorus intake affects the retention of calcium adversely. Many of these studies were carried out in animals (7-9). Studies carried out in man have shown that phosphorus given in large amounts has an effect on parathyroid hormone (PTH) levels (10,11). Extensive investigations carried out in man in this Research Unit have shown that increasing the phosphorus intake by a factor of approximately 2.5, i.e., from 800 to 2000 mg per day, during different intake levels of calcium, resulted in a significant decrease of the urinary calcium, the fecal calcium increased slightly but not significantly and the calcium balance was not affected, and in fact, it became even slightly more positive in some cases during the high phosphorus intake (12). The decrease of the urinary calcium, observed during the high phosphorus intake, is not an undesirable effect on calcium metabolism. Also, the intestinal absorption of calcium was not decreased during the high phosphorus intake although one might expect this to occur in view of the high phosphorus content in the intestine and the formation of calcium-phosphate precipitates. The observations made in studies in our Research Unit lead one to conclude that the dietary Ca/P ratio does not appear to play an important role, a viewpoint which has also been expressed by the World Health Organization (13) and by other investigators (14). Other investigators have also reported that phosphorus causes a decrease in urinary calcium (15,16) and promotes healing of skeletal fractures (15).

A representative example of the effect of phosphorus given as sodium glycerophosphate is shown on Table I. Increasing the phosphorus intake from 800 mg to 2000 mg per day during a normal calcium intake of approximately 850 mg per day resulted in a decrease of the urinary calcium from 301 mg to 179 mg per day, the stool calcium increased slightly and the calcium balance became somewhat more positive.

Magnesium. The recommended dietary allowance, RDA, for magnesium is

Table I.   Effect of Certain Minerals on the Calcium Balance and on Calcium Absorption

STUDY	MINERAL INTAKE, mg/day	STUDY DAYS	CALCIUM, mg/day				47Ca ABSORPTION, % dose
			INTAKE	URINE	STOOL	BALANCE	
Control-Phosphorus	800	30	852	301	521	+ 30	49
High Phosphorus+	2000	36	855	179	560	+116	43
Control-Magnesium	245	30	866	134	685	+ 47	42
High Magnesium++	800	36	868	129	705	+ 34	40
Control-Fluoride	4	32	1772	152	1509	+111	19
High Fluoride*	48	94	1768	126	1541	+101	22
Control-Zinc	15	30	230	93	299	-162	61
High Zinc**	155	94	233	45	350	-163	39

+ High phosphorus intake due to addition of sodium glycerophosphate.
++ High magnesium intake due to addition of magnesium oxide.
* High fluoride intake due to addition of sodium fluoride.
** High zinc intake due to addition of zinc sulfate during a low calcium intake.

300 mg per day for women and 350 mg per day for men (1). Studies
carried out in this Research Unit have shown that increasing the
magnesium intake from 245 mg per day to 800 mg per day by adding
magnesium as the oxide to the constant diet did not result in a
change of urinary or fecal calcium excretion nor of the calcium balance (See Table I). Calcium absorption studies, using [47]Ca as the
tracer, have shown that there was no change of the intestinal absorption of calcium during the high magnesium intake (17). The lack
of change of the calcium balance during a high magnesium intake is
in agreement with observations made by others (18). Increased magnesium intake has been reported by others to increase urinary calcium and this increase could be suppressed by the addition of phosphate (19-21).

Fluoride. Fluoride is contained in water and in the diet (22,23).
Also, fluoride stabilizes the bone crystal (24) and has been shown
to decrease bone resorption (25). For these reasons, fluoride has
been recommended for the treatment for osteoporosis (26-28). The
importance of the role of fluoride on maintaining the normal bone
structure is also indicated by a survey which showed that the incidence of osteoporosis was lower in naturally high fluoride areas
than in areas where the fluoride content of the drinking water was
low (29). This observation would indicate that the intake of certain amounts of fluoride throughout life may be necessary for the
maintenance of the normal skeleton, and that fluoride may protect
the skeleton from the development of bone loss which leads to osteoporosis with advancing age, particularly in females.

In extensive studies of the effect of fluoride in patients with
osteoporosis carried out in this Research Unit, a therapeutic daily
dose of 45 mg fluoride per day was used as sodium fluoride. The
main effect of fluoride on calcium metabolism was a decrease of the
urinary calcium, while the fecal calcium did not change and the calcium balance also remained unchanged. Also, the intestinal absorption of calcium remained unchanged during the high fluoride intake
(Table I). The decrease in urinary calcium, induced by sodium
fluoride, may be due to decreased bone resorption, a very desirable
effect for patients with osteoporosis.

Zinc. A great deal of information is available on the effect of calcium on zinc metabolism, particularly in animals (30,31), but only a
few studies have been carried out in man (32,33). On the other hand,
to our knowledge, no studies have been reported on the effect of
zinc on calcium metabolism in man. In studies carried out in this
Research Unit, it has been demonstrated that large amounts of zinc
significantly decrease the intestinal absorption of calcium if the
zinc supplements are given during a low calcium intake (34).
Table I shows that the addition of 140 mg of zinc as zinc sulfate
per day, during a low calcium intake of 230 mg/day, decreased the
urinary calcium, increased the fecal calcium and the calcium balance
remained unchanged. In nine patients studied during a low calcium
intake, the intestinal absorption of calcium decreased significantly
during the high zinc intake, the absorption values averaging
$61 \pm 6.6\%$ in the control study and $39 \pm 4.6\%$ during zinc supplementation. However, the decrease in calcium absorption was abolished

when the calcium intake was increased to 800 mg per day during the
high zinc intake. The studies have also shown that the decrease in
calcium absorption during the high zinc intake did not depend on the
form of zinc. This was demonstrated in a study which yielded similar
results when zinc was given as zinc sulfate or zinc gluconate to the
same subject. Other investigators have reported that excess zinc intake decreased bone ash (35) and increased bone resorption (36).
These observations were explained by decreased intestinal absorption
of calcium and by subsequent increased stimulation of the secretion
of PTH and removal of calcium from bone.

Protein Intake. A high protein intake has been reported to induce
calcium loss in animals and in man. It should, however, be pointed
out that the source of protein appears to play a major role in induc-
ing this effect. While purified proteins have been shown to increase
calcium loss (37-39), the use of red meat as the source of protein
(40,41) did not lead to an increase in the excretion of calcium, and
therefore, it did not result in calcium loss. Table II shows ex-
amples of the lack of effect of a high protein diet given as meat on
urinary calcium. In these long term studies of 132 and 102 days,
respectively, the urinary calcium did not change, the fecal calcium
also remained unchanged as did the calcium balance. ^{47}Ca absorption
studies (not shown in Table II) have revealed that a high protein
intake of 2 gm/kg body weight did not alter the intestinal absorption
of calcium compared to absorption values obtained during the intake
of 1 gm protein per kilogram body weight. The difference observed in
high protein studies using purified proteins and in our studies in
man using meat as the source of protein appears to be due to the high
phosphorus content of meat. This is pointed out because phosphorus
has been shown to decrease the urinary calcium excretion (12,15,16),
and the phosphorus content of meat may counteract any increase of the
urinary calcium which may have been induced by a high protein intake
given as meat.

Table II. Effect of a High Protein (Meat) Diet on the Calcium Balance

Patient	Protein Intake gm/kg	Study Days	Calcium, mg/day			
			Intake	Urine	Stool	Balance
1	1	64	848	187	707	−46
	2	132	830	156	719	−45
2	1	30	825	205	606	+14
	2	102	899	202	659	+36

Effects of Drugs on Calcium Metabolism

Several drugs are known to affect calcium metabolism and to induce
calcium loss (42). Amongst these are the use of glucocorticoids and
thyroid extract. The use of both of these drugs can lead to the development of osteoporosis. It is less well known that several other

commonly used drugs can cause increased excretion of calcium and may
thereby lead to calcium loss.

Patients who previously had the diagnosis of pulmonary tubercu-
losis, usually receive for many years the drug Isoniazid (INH) in
order to prevent the recurrence of this disease. Studies carried
out in this Research Unit have shown that the urinary calcium excretion of patients who received this drug for prolonged periods of
time, i.e., for several years, is very high and that the calcium balance is quite negative. Another commonly used drug is the anti-
biotic Tetracycline, which has also been shown to cause a significant
increase in urinary calcium after its long-term use, namely after its
use for several years. Studies carried out in this Research Unit
have also shown that the short-term use of Tetracycline can also increase the urinary calcium excretion. The diuretic Furosemide
(Lasix) which is widely used, particularly for patients with hypertension or for those who suffer from congestive heart failure, also
results in an increase of urinary calcium excretion. This increase
is associated with an increase of the excretion of sodium in urine.

Aluminum. Antacids are commonly used by the public for various
gastrointestinal disturbances, and because of their widespread use,
the effects of aluminum-containing antacids on calcium metabolism are
described in more detail. The calcium loss induced by these antacids
can accentuate or accelerate demineralization of the skeleton which
may be due to other causes such as aging or hormonal deficiency.
Studies carried out in our Research Unit (43,44) have shown that
even small doses of aluminum-containing antacids induce calcium loss
(Table III). This is particularly true for larger doses of these
antacids (45). The primary effect of these antacids in relation to
mineral metabolism is inhibition of the intestinal absorption of
phosphorus by aluminum due to complexation of phosphorus by aluminum
in the intestine, resulting in phosphorus depletion. A secondary
effect of phosphorus depletion is an increase in calcium loss due to
increased urinary calcium excretion. However, there is also an increase in fecal calcium, the calcium losses by both routes resulting
in a negative calcium balance.

Table III shows data obtained in studies which were carried out
in this Research Unit using relatively small doses of aluminum-
containing antacids. These studies have shown that even these doses
have a similar adverse effect as large doses in causing phosphorus
depletion and a secondary increase in urinary calcium. A daily dose
of 90 ml Maalox per day increased both the urinary and fecal calcium
excretion and the calcium balance became more negative. When the
dose of Maalox was increased to therapeutic levels which are commonly used for patients with acute peptic ulcer or gastritis, the urinary calcium markedly increased from 82 to 421 mg per day, the fecal
calcium remained unchanged and the calcium balance became highly
negative, -502 mg per day. Studies of the intestinal absorption of
calcium, determined with tracer doses of ^{47}Ca, have shown that the
absorption of calcium did not change during the intake of aluminum-
containing antacids. The increase in urinary calcium appears to be
due to increased bone resorption. Extensive calcium loss, evidenced
on x-rays of the skeleton, has been seen in several of our patients
who have taken on their own, commonly used aluminum-containing
antacids for many years.

Table III. Effect of Small and Large Doses of Aluminum-Containing Antacids on Calcium Metabolism

Study	Study Days	Calcium, mg/day			
		Intake	Urine	Stool	Balance
Control	36	229	139	113	− 23
Antacid[+]	18	259	186	157	− 84
Control	18	253	82	348	−177
Antacid*	12	279	421	360	−502

[+] Dose of Maalox = 90 ml/day
* Dose of Maalox = 450 ml/day

Literature Cited

1. "Recommended Dietary Allowances," National Academy of Sciences, 1980, 9th ed.
2. Heaney, R. P. *J. Lab. Clin. Med.* 1982, 100, 309–12.
3. Spencer, H.; Menczel, J.; Lewin, I.; Samachson, J. *Am. J. Med.* 1964, 37, 232–34.
4. Spencer, H.; Kramer, L.; Osis, D. *Am. J. Clin. Nutr.* 1982, 36, 776–87.
5. Spencer, H.; Friedland, J. A.; Ferguson, V. In "Biological Mineralization"; Zipkin, I., Ed.; John Wiley & Sons, Inc.: New York, 1973; pp. 689–727.
6. Willis, J. B. *Anal. Chem.* 1961, 33, 556–9.
7. Clark, I. *Amer. J. Physiol.* 1969, 217, 865–70.
8. Draper, H. H.; Sie, T-L.; Bergan, J. C. *J. Nutr.* 1972, 102, 1133–42.
9. Laflamme, G. H.; Jowsey, J. *J. Clin. Invest.* 1972, 51, 2834–40.
10. Reiss, E.; Canterbury, J. M.; Bercovitz, M. A.; Kaplan, L. *J. Clin. Invest.* 1970, 49, 2146–9.
11. Bell, R. R.; Draper, H. H.; Tzeng, D. Y. M.; Shin, H. K.; Schmidt, G. R. *J. Nutr.* 1977, 107, 42–50.
12. Spencer, H.; Kramer, L.; Osis, D.; Norris, C. *J. Nutr.* 1978, 108, 447–57.
13. "Calcium Requirements"; WORLD HEALTH ORGANIZATION TECHNICAL REPORT SERIES No. 230, WHO: Geneva, 1962; pp. 16–18.
14. Hegsted, D. M. *Fed. Proc.* 1967, 26, pp. 1747–54.
15. Goldsmith, R. S.; Woodhouse, C. F.; Ingbar, S. H.; Segal, O. *Lancet* 1967, 1, 687–90.
16. Zemel, M. B.; Linkswiler, H. M. *J.Nutr.* 1981, 111, 315–24.
17. Spencer, H.; Lesniak, M.; Gatza, C. A.; Kramer, L.; Norris, C.; Coffey, J. In "Trace Substances in Environmental Health-XII"; Hemphill, D. D., Ed.; University of Missouri: Columbia, 1978, pp. 241–47.

18. Schwartz, R.; Woodcock, N. A.; Blakely, J. D.; MacKellar, I. Am. J. Clin. Nutr. 1973, 26, 519-23.
19. Edwards, N.; Russell, R.; Hodgkinson, A. Brit. J. Urol. 1965, 37, 390-8.
20. Heaton, F. W.; Hodgkinson, A.; Rose, G. A. Clin. Sci. 1964, 27, 31-40.
21. Tibbetts, D. M.; Aub, J. C. J. Clin. Invest. 1937, 16, 491-501.
22. Osis, D.; Kramer, L.; Wiatrowski, E.; Spencer, H. J. Nutr. 1974, 104, 1313-18.
23. Kramer, L.; Osis, D.; Wiatrowski, E.; Spencer, H. Am. J. Clin. Nutr. 1974, 27, 590-4.
24. Zipkin, I.; Posner, A. S.; Eanes, E. D. Biochim et Biophys Acta. 1962, 59, 255-8.
25. Jowsey, J.; Kelly, P. J. Mayo. Clin. Proc. 1968, 43, 435-43.
26. Rich, C.; Ensinck, J.; Ivanovich, P. J. Clin. Invest. 1964, 43, 545-56.
27. Spencer, H.; Lewin, I.; Osis, D.; Samachson, J. Am. J. Med. 1970, 49, 814-22.
28. Jowsey, J.; Riggs, B. L.; Kelly, P. J.; Hoffman, D. L. Am. J. Med. 1972, 53, 43-9.
29. Bernstein, D.; Sadowsky, N.; Hegsted, D. M.; Guri, C.; Stare, F. J.A.M.A. 1966, 198, 499-504.
30. Oberleas, D.; Muhrer, M. E.; O'Dell, B. L. J. Animal Sci. 1962, 21, 57-61.
31. O'Dell, B. L.; Savage, J. E. Proc. Soc. Exp. Biol. Med. 1960, 103, pp. 304-6.
32. Pecoud, A.; Donzel, P.; Schelling, J. L. Clin. Pharm. Therap. 1975, 17, 469-74.
33. Spencer, H.; Kramer, L.; Osis, D. In "Clinical, Biochemical and Nutritional Aspects of Trace Elements"; Prasad, A. S., Ed.; Alan R. Liss: New York, 1982, pp. 103-115.
34. Spencer, H.; Osis, D.; Kramer, L. Am. J. Clin. Nutr. 1977, 30, 611.
35. Sadasivan, V. Biochim. J. 1951, 48, 527-30.
36. Ferguson, H. W.; Leaver, A. G. Calcif. Tissue Res. 1972, 8, 265.
37. Bell, R. R.; Engelman, D. T.; Sie, T.; Draper, H. H. J. Nutr. 1975, 105, 475-83.
38. Hegsted, D. M.; Linkswiler, H. M. J. Nutr. 1981, 111, 120-7.
39. Linkswiler, H. M.; Zemel, M. B.; Hegsted, M.; Schuette, S. Fed. Proc. 1981, 40, pp. 2429-33.
40. Spencer, H.; Kramer, L.; Osis, D.; Norris, C. Am. J. Clin. Nutr 1978, 31, 2167-80.
41. Spencer, H.; Kramer, L; DeBartolo, M.; Norris, C.; Osis, D.; Am. J. Clin. Nutr. 1983, 37, 924-9.
42. Spencer, H.; Kramer, L.; Osis, D. Am. J. Clin. Nutr. 1982, 36, 776-87.
43. Spencer, H.; Kramer, L.; Norris, C.; Osis, D. Am. J. Clin. Nutr 1982, 36, 32-40.
44. Spencer, H.; Kramer, L. Arch. Int. Med. 1983, 143, 657-9.
45. Lotz, M.; Zisman, E.; Bartter, F. C. N. Engl. J. Med. 1968, 278, 409-15.

RECEIVED January 15, 1985

Influence of Zinc on Calcium Bioavailability

ADEN C. MAGEE

Department of Food, Nutrition, and Food Service Management, School of Home Economics, University of North Carolina, Greensboro, NC 27412

High levels of dietary zinc were associated with marked decreases in bone calcium deposition and in the apparent retention of calcium in male weanling albino rats. Marked increases in fecal calcium levels were also observed in the zinc-fed rats. Excessive dietary zinc was associated with a shifting of phosphorus excretion from the urine to the feces. This resulted in an increase in fecal phosphorus and provided an environmental condition which would increase the possibility of the formation of insoluble calcium phosphate salts and a subsequent decrease in calcium bioavailability. The adverse effect of high dietary zinc on calcium status in young rats could be alleviated and/or reversed with calcium supplements.

Most of the research on the influence of zinc on calcium bioavailability has been in connection with zinc toxicity or the effects of high levels of dietary zinc on various animal systems. Such studies and/or investigations have been conducted on a variety of animal species and humans, but those studies which have revealed a possible effect of zinc on calcium bioavailability have generally involved the lamb, pig, and rat. This paper will be primarily a discussion of the effects of high levels of dietary zinc on calcium status in the rat. The effect of zinc on phosphorus status, however, has been included because there is the possibility that the effect of zinc on calcium bioavailability may be dependent upon the phosphorus status of the system.

The first experimental indication of an adverse effect of excess dietary zinc on animals was revealed by Sutton and Nelson ([1]) in 1937. These researchers observed decreased growth, anemia, and reproductive failures in rats fed diets containing 0.5% to 1% zinc. Following this initial report, several studies were conducted by various researchers, but most of the studies pertaining to zinc toxicity between the late 1930's and the early 1950's primarily focused on the effects of zinc toxicity on growth and on copper and iron metabolism. During 1951 and 1952 a series of reports by

Sadasivan (2-4) revealed that 0.5% and 1% levels of dietary zinc increased the fecal and urinary excretion of phosphorus and caused an impairment in the development and the mineralization of the bones of young rats. Sadasivan suggested that the impairment in bone development could be attributed to an antagonistic effect of zinc upon normal bone deposition of calcium and phosphorus. In 1958 Whiting and Bezeau (5) reported that dietary zinc increased the retention of calcium in pigs, but had no influence on phosphorus excretion. Thompson, Hansard, and Bell (6), however, reported the following year that high levels of dietary zinc were associated with decreases in the retentions of both calcium and phosphorus in lambs.

My interest in zinc was stimulated in the late 1950's, and for several years I was one of several researchers looking at various aspects of zinc toxicosis in connection with a variety of biological parameters. Beginning in 1961 one area of concentration was the effects of high levels of zinc on calcium and phosphorus status in young rats, and some of the results of studies (7,8) conducted in our laboratory form the major portion of this paper.

. All of the studies were conducted with weanling, male albino rats of the Sprague-Dawley strain (Holtzman company). The basal diet used for these studies consisted of casein, starch, vegetable oil, vitamin and mineral mixtures, and cellulose. The Wesson Modification of the Osborne-Mendel mineral mixture was used in all studies. This mineral mixture contained no zinc, but it was adequate in the other minerals required by the rat. Most of the non-zinc-supplemented diets used in the various experiments contained approximately 7 ppm zinc. The level of mineral mixture used in the basal diets was 4%, and based on the chemical composition of the mixture, the basal diets contained approximately 0.57% calcium and 0.41% phosphorus.

In one of the first experiments, weanling rats were fed 0.25%, 0.5%, and 0.75% zinc for 3 weeks. Animals were sacrificed at weekly intervals, and the effects of these levels of zinc on weight gains and on bone calcium and phosphorus levels are shown in Table I. Although none of the levels of dietary zinc had any effect on weight gains at the end of the first week, levels of 0.5% and 0.75% zinc were associated with decreases in bone calcium and phosphorus levels at the end of the first week. Analysis of the data indicated that weight gains of young rats fed 0.75% zinc were markedly lower than the weight gains of the controls at the end of the second week.

Table II shows the effects of varying dietary levels of zinc on weight gains and on bone calcium and phosphorus levels of young rats at the end of a 4-week experiment. Increases in dietary zinc were associated with significant linear decreases in bone calcium and phosphorus deposition. The bones taken from animals at the time of sacrifice and used for the mineral analyses were very soft in nature and could be easily squeezed with the fingers.

A second step in the investigations involved the addition of calcium and phosphorus supplements to young rats fed 0.75% zinc, and Table III shows some of the results when 0.4% levels of calcium and phosphorus were used as supplements. The high level of dietary zinc again resulted in marked decreases in weight gains and in bone calcium and phosphorus levels. Supplementing the high zinc diets with 0.4% calcium resulted in bone calcium levels which were sig-

Table I.   Effects of Zinc on Growth and Bone Mineralization of
Young Rats

Weeks on Experiment	Zinc Supplement	Weight Gain[1]	Bone Constituent[2] Ca	Bone Constituent[2] P
	%	gm	mg/gm dry weight	
1	None	39 + 1	141 + 4	92 + 1
	0.25	33 + 1	132 + 2	87 + 2
	0.50	31 + 1	113 + 4	85 + 4
	0.75	27 + 1	116 + 4	84 + 2
2	None	87 + 2	151 + 1	92 + 1
	0.25	76 + 4	143 + 5	90 + 1
	0.50	67 + 2	122 + 8	84 + 2
	0.75	61 + 2	114 + 3	81 + 2
3	None	134 + 3	156 + 6	94 + 1
	0.25	121 + 4	145 + 6	89 + 1
	0.50	107 + 2	142 + 7	89 + 1
	0.75	88 + 4	126 + 7	84 + 1

Note: Sprague-Dawley rats averaging 49 gm in weight initially.
Each value is the mean + SEM.
[1]Mean of 16, 12, and 8 animals for weeks 1, 2, and 3, respectively.
[2]Means of 4 animals.
Source: Reproduced with permission from Ref. 7. Copyright 1964
J. Nutr., American Institute of Nutrition.

Table II. Effects of Zinc on Growth and Bone Mineralization of
Young Rats

Zinc Supplement	Weight Gain at 4 Weeks	Bone Constituent Ca	Bone Constituent P
%	gm	mg/gm dry weight	
None	188 + 8	146 + 8	72 + 2
0.25	197 + 6	144 + 6	70 + 2
0.50	161 + 6	131 +10	66 + 1
0.75	113 + 9	107 + 9	61 + 3

Note: Sprague-Dawley rats averaging 53 gm in weight initially.
Each value is the mean of 6 animals + SEM.
Source: Reproduced with permission from Ref. 7. Copyright 1964
J. Nutr., American Institute of Nutrition.

Table III. Effects of Calcium, Phosphorus, and Zinc Supplements
on Growth and Bone Mineralization of Young Rats

Supplement			4 Weeks	Bone Constituent[2]	
Zn	Ca	P	Weight Gain[1]	Ca	P
%			gm	mg/gm dry weight	
0	0	0	182 ± 8	163 ± 9	87 ± 4
0.75	0	0	97 ± 8	111 ± 5	77 ± 6
0.75	0.4	0	126 ± 9	134 ± 8	81 ± 1
0.75	0	0.4	106 ± 9	109 ± 8	75 ± 3
0.75	0.4	0.4	154 ± 5	158 ± 8	88 ± 5

Note: Sprague-Dawley rats averaging 49 gm in weight initially.
[1]Each figure is the mean of 6 animals ± SEM.
[2]Each figure is the mean of 4 animals ± SEM.
Source: Reproduced with permission from Ref. 7. Copyright 1964
J. Nutr., American Institute of Nutrition.

nificantly higher than those of rats fed only zinc. Although supplements of 0.4% phosphorus did not prevent the marked decreases in bone calcium associated with the addition of zinc to the diet, the combination of phosphorus with calcium resulted in increases in bone calcium levels greater than those observed in the zinc-fed rats receiving only calcium supplements. The bone calcium levels of the zinc-fed rats supplemented with both calcium and phosphorus were essentially the same as the bone calcium levels of the control animals. Additional calcium appeared to improve bone phosphorus levels of rats fed the high zinc diet, but phosphorus supplements appeared to have little effect on bone phosphorus levels, except in the presence of additional calcium.

Table IV shows data from another experiment in which the levels of calcium and phosphorus supplements were 0.4%, 0.8%, and 1.2%. Supplements of calcium resulted in significant improvements in bone calcium levels of zinc-fed rats, whereas phosphorus supplements had little effect on bone calcium levels in zinc-fed rats. Increasing the calcium supplement from 0.4% to 0.8% resulted in additional increases in bone calcium deposition, and a level of 1.2% added calcium resulted in the same increase in bone calcium levels of rats fed 0.75% zinc as did the 0.8% calcium supplement. Mean bone calcium values of zinc-fed rats supplemented with either 0.4% calcium and phosphorus or 0.8% calcium and phosphorus were essentially the same as mean bone calcium values of zinc-fed rats supplemented with either 0.8% or 1.2% calcium. A combination supplement of 1.2% calcium and phosphorus was found to completely alleviate the decrease in bone calcium deposition associated with the feeding of a 0.75% level of zinc. Calcium supplements partially alleviated the decrease in bone phosphorus level associated with the feeding of the

Table IV. Effects of Calcium, Phosphorus, and Zinc Supplements
on Growth and Bone Mineralization of Young Rats

Supplement			4 Weeks	Bone Constituent[2]	
Zn	Ca	P	Weight Gain[1]	Ca	P
%			gm	mg/gm dry weight	
0	0	0	148 + 5	163 + 5	83 + 2
0.75	0	0	104 + 9	116 + 6	73 + 1
0.75	0.4	0	124 + 4	132 + 5	79 + 3
0.75	0	0.4	113 + 3	118 + 4	74 + 2
0.75	0.4	0.4	131 + 3	147 + 7	75 + 2
0.75	0.8	0	131 + 2	149 + 4	80 + 1
0.75	0	0.8	116 + 5	125 + 3	70 + 4
0.75	0.8	0.8	132 + 6	147 + 4	80 + 2
0.75	1.2	0	120 + 4	147 + 3	78 + 1
0.75	0	1.2	119 + 3	126 + 4	73 + 2
0.75	1.2	1.2	115 + 4	157 + 4	80 + 2

Note: Sprague-Dawley rats averaging 47 gm in weight initially.
[1]Each figure is the mean of 6 animals + SEM.
[2]Each figure is the mean of 4 animals + SEM.
Source: Reproduced with permission from Ref. 7. Copyright 1964
J. Nutr., American Institute of Nutrition.

high level of zinc, but phosphorus supplements had no apparent beneficial effect on bone phosphorus deposition in the zinc-fed rats. A combination of either 0.8% calcium and phosphorus or 1.2% calcium and phosphorus was associated with bone phosphorus levels approaching those observed in the non-zinc supplemented rats.

The effect of zinc toxicity on the calcium balance of young rats is presented in Table V. Marked increases in the fecal excretion of calcium was noted in the zinc-fed rats, and the overall effect of zinc toxicity was a substantial decrease in the apparent retention of calcium. The effects of zinc on calcium retention were noted as early as one week of the experimental period.

Table VI shows the effect of 0.75% dietary zinc on the phosphorus balance in young rats. A decrease in the apparent retention of phosphorus was noted in the zinc-fed rats as early as the end of the first week. Possibly a more significant observation was the apparent movement of phosphorus excretion from the urine, the normal pathway for phosphorus excretion, to the feces in rats fed the high zinc diet. Such a shifting of the phosphorus excretion to the fecal pathway in animals fed a high zinc diet should result in an increase

Table V. Effect of Zinc Toxicity on Calcium Retention in Young Rats

Weeks on Experiment	Zinc Supplement	Ca Intake	Fecal Ca	Urinary Ca	Ca Retention
	%	mg/day	mg/day	mg/day	mg/day
1	0	48.5	16.5	0.1	31.9
	0.75	45.7	31.1	0.3	14.3
2	0	56.2	21.1	0.1	35.0
	0.75	51.3	45.0	0.2	6.1
3	0	58.5	27.4	0.2	30.9
	0.75	54.2	40.3	0.3	13.6
4	0	67.4	34.1	0.3	33.0
	0.75	57.9	48.5	0.3	9.1

Note: Each figure is the mean of 4 animals.
Source: Reprinted with permission from Ref. 7. Copyright 1964
J. Nutr., American Institute of Nutrition.

Table VI. Effect of Zinc Toxicity on Phosphorus Retention in Young Rats

Weeks on Experiment	Zinc Supplement	P Intake	Fecal Ca	Urinary P	P Retention
	%	mg/day	mg/day	mg/day	mg/day
1	0	63.5	8.0	14.8	40.7
	0.75	58.0	21.0	7.5	29.5
2	0	73.7	8.6	20.4	44.7
	0.75	65.1	28.6	14.8	21.7
3	0	76.9	11.7	23.6	41.6
	0.75	68.8	21.6	14.1	33.1
4	0	88.4	14.9	26.7	46.8
	0.75	73.5	26.3	14.4	32.8

Note: Each figure is the mean of 4 animals.
Source: Reproduced with permission from Ref. 7. Copyright 1964
J. Nutr., American Institute of Nutrition.

in the amount of phosphorus in the lower region of the gastrointestinal tract and subsequently provide a favorable environment
for the formation of insoluble calcium-phosphate salts. If this
does occur, as these data suggest, the subsequent loss of calcium
from the system or the decrease in the bioavailability due to zinc
toxicity may be the result of phosphorus status in the system. Zinc
supplementation may exert a major influence in facilitating a shift
in the phosphorus excretion pathway.

Table VII shows the calcium balance of zinc-fed and non-zinc-
fed rats supplemented with 0.8% calcium and/or phosphorus. Marked
increases in fecal calcium and corresponding decreases in apparent
calcium retentions in the zinc-fed rats could be reversed with
calcium supplementation. Phosphorus supplements appeared to be
associated with increases in calcium retention in the absence of
zinc, but decreases in calcium retention in the presence of zinc
without calcium supplementation. Decreases in fecal calcium were
noted in animals fed calcium supplements in the presence of phosphorus or zinc. High levels of zinc were associated with increases
in fecal calcium excretion in the absence of extra calcium or in the
presence of extra phosphorus. Calcium supplementation was generally
associated with a decrease in the urinary excretion of calcium,
while zinc and phosphorus supplements were generally associated with
an increase in urinary calcium excretion.

Table VII.   Effects of Calcium, Phosphorus, and Zinc Supplements
             on Calcium Retention in Young Rats

Diet	Ca Intake	Fecal Ca	Urinary Ca	Ca Retention
	mg/day	percentage of intake		
Basal	66.9	58.7	0.7	40.6
+ 0.8% Ca	288.1	49.9	0.4	49.7
+ 0.8% P	82.0	43.0	1.9	55.1
+ 0.8% Ca + 0.8% P	292.2	30.7	0.7	68.4
+ 0.75% Zn	61.4	91.0	1.3	8.1
+ 0.75% Zn + 0.8% Ca	208.2	58.1	0.2	41.5
+ 0.75% Zn + 0.8% P	79.4	107.0	2.4	-9.4
+ 0.75% Zn + 0.8% Ca + 0.8% P	232.6	56.8	0.5	42.7

Note: Each figure is the mean of 4 animals.
Source: Reproduced with permission from Ref. 8. Copyright 1964
J. Nutr., American Institute of Nutrition.

Table VIII shows the phosphorus balance of zinc-fed and non-
zinc-fed rats supplemented with 0.8% calcium and/or phosphorus. A
dietary level of 0.75% zinc was associated with decreases in the
apparent retention of phosphorus, and these data support the
findings of Sadasivan (2-4). Calcium supplements were associated
with slight increases in the retention of phosphorus, while phos-

phorus supplements were associated with decreases in phosphorus
retention. Calcium and zinc supplements were associated with increases in fecal phosphorus excretion, and phosphorus supplements
were associated with increases in the urinary excretion of phosphorus. Interpretation of significant Ca x Zn and Ca x P x Zn
interactions suggested a combination of calcium plus zinc resulted
in additional increases in fecal phosphorus levels, while the
presence of phosphorus appeared to moderate the effects of calcium
or zinc on fecal phosphorus excretion. A significant P x Zn interaction suggested that zinc resulted in an increase in urinary phosphorus excretion in this particular experiment in the absence of
phosphorus and a decrease in urine phosphorus excretion in the
presence of phosphorus.

Table VIII. Effects of Calcium, Phosphorus, and Zinc Supplements
on Phosphorus Retention in Young Rats

Diet	P Intake	Fecal P	Urinary P	P Retention
	mg/day	percentage of intake		
Basal	71.4	20.0	28.3	51.7
+ 0.8% Ca	75.2	57.5	3.6	51.7
+ 0.8% P	176.4	12.4	55.1	32.4
+ 0.8% Ca + 0.8% P	190.0	26.9	38.6	34.5
+ 0.75% Zn	60.7	57.2	28.6	14.2
+ 0.75% Zn + 0.8% Ca	74.1	60.2	12.5	27.6
+ 0.75% Zn + 0.8% P	164.4	29.3	50.5	20.2
+ 0.75% Zn + 0.8% Ca + 0.8% P	187.4	38.4	32.5	29.1

Note: Each figure is the mean of 4 animals.
Source: Reproduced with permission from Ref. 8. Copyright 1964
J. Nutr., American Institute of Nutrition.

     The data presented in this paper indicate that excess levels
(0.75%) of dietary zinc result in decreases in the bioavailability
of calcium and phosphorus in rats and interfere with normal bone
mineralization. High dietary levels of calcium or zinc appeared to
cause a shift in the excretion of phosphorus from the urine to the
feces, while the presence of extra phosphorus tended to keep the
pathway of phosphorus excretion via the urine. The presence of
large amounts of phosphorus in the intestinal tract due to high
intakes of zinc would increase the possibility of the formation of
insoluble phosphate salts with various cations, including calcium,
which may be present. A shift in phosphorus excretion from the
feces to the urine, however, could result in an environmental condition within the system which would tend to increase the bioavailability of cations to the animal. The adverse effect of zinc toxicity on calcium and phosphorus status of young rats could be alleviated with calcium and/or phosphorus supplements.

Acknowledgments

Experiments reviewed in this paper were conducted in connection with projects financed by the North Carolina Agricultural Research Service, North Carolina State University, Raleigh, North Carolina. Paper No. 9305 of the Journal Series of the North Carolina Agricultural Research Service. The use of trade names in this publication does not imply endorsement by the North Carolina Agricultural Research Service of products named, nor criticism of similar ones not mentioned. The author is indebted to Mrs. Betty Weaver for assistance with the preparation of this paper.

Literature Cited

1. Sutton, W. R.; Nelson, V. E. Proc. Soc. Exptl. Biol. Med. 1937, 36, 211–213.
2. Sadasivan, V. Biochem. J. 1951, 48, 527–530.
3. Sadasivan, V. Biochem. J. 1951, 49, 186–191.
4. Sadasivan, V. Biochem. J. 1952, 52, 452–455.
5. Whiting, F.; Bezeau, L. M. Canadian J. Animal Sci., 1958, 38, 109.
6. Thompson, A.; Hansard, S. L.; Bell, M. C. J. Animal Sci. 1959, 18, 187.
7. Stewart, A. K.; Magee, A. C. J. Nutrition 1964, 82, 287–295.
8. Magee, A. C.; Fu, S. C. Nutr. Reports Inter. 1979, 19, 343–351.

RECEIVED October 15, 1984

Effect of Dietary Fat and Fiber on Calcium Bioavailability

CONSTANCE KIES

Department of Human Nutrition and Food Service Management, University of Nebraska, Lincoln, NE 68583

Among the many dietary factors which may influence the absorption of calcium from the intestines are kind and level of dietary fiber and kind and level of dietary fat. Over the last 40 years, considerable amounts of evidence have been gathered from epidemiological and controlled laboratory studies concerning interactions between dietary fiber intake and bioavailability of calcium. Apparently certain kinds of dietary fiber fed under certain conditions can have an adverse effect on the intestinal absorption of calcium. Whether or not a specific kind of dietary fiber has an effect on calcium absorption is dependent upon the chemical-physical properties of the particular fiber, food or diet in which it is included, the level fed, and physiological characteristics of the individual eating it. The role of dietary fat as a factor influencing the intestinal absorption of calcium has been less extensively investigated. Fatty diarrhea in which the usual efficient fat absorption is deranged also has a pronounced adverse effect on the absorption of calcium. Under some conditions, calcium bioavailability appears to decrease as chain length of fatty acids increase and as the degree of unsaturation decreases. Results of studies designed to investigate effects of level of dietary fat on calcium absorption have given mixed results.

High incidences of atherosclerosis/coronary heart disease and cancer in the United States have led to programs encouraging Americans to change their eating patterns relative to consumption of dietary fiber and fat (1). However, increased interest in possible dietary or physiological factors involved in maintenance of bone health throughout life has developed because of the high incidence of osteoporosis among elderly Americans (2,3). The role of calcium bioavailability in bone mineral homeostasis has not been fully

0097–6156/85/0275–0175$06.00/0

defined. However, it would appear advantageous to define all
dietary contributions to calcium nutritional status. In particular,
all consequences of suggested dietary changes relative to reducing
incidence of one type of pathological condition should be investigated so that additional, concurrent recommendations may be made if
found to be desirable.

Dietary Fiber and Calcium Bioavailability

More than 40 years ago, calcium absorption from brown (whole wheat)
bread which was fed to human subjects was found to be poorer than
was that when white (extracted wheat flour) was fed (4,5). Since
then, many studies have sought to define the extent of inhibition
of calcium intestinal bioavailability by various forms of dietary
fiber with mixed results and conclusions (6-18).
     Serum calcium levels were depressed with the feeding of 18 to
100 g of wheat bran to human subjects in the study by Heaton and
Pomare (10) but were unaffected by the feeding of approximately 20 g
of wheat fiber in the study by Jenkins et al. (19). Since blood
serum calcium levels tend to be resistant to change except under
conditions of severe calcium deficiency, measurement of blood serum
calcium levels is probably not sufficiently sensitive to indicate
change in calcium nutritional status in short-term feeding studies.
     Most of the forementioned studies which examined the influence
of various dietary fiber on the bioavailability of calcium by human
subjects have depended upon the comparative measurements of calcium
content of diets and calcium contents of stools and urine. As
reviewed by Allen (3), calcium balance studies have distinct limitations relative to accuracy and precision. However, their ease of
application and cost, laboratory equipment requirements, and real
(or perceived) safety in comparison to available radioactive or
stable isotope methods continue to make their use popular. In
calcium balance studies, calcium absorption is assumed to be the
difference between calcium excretion in the feces and calcium
intake. Usually this is expressed as a percent of the calcium
intake. This method assumes that all fecal calcium loss is
unabsorbed dietary calcium which is, of course, untrue since
appreciable amounts of calcium from the body are lost via the
intestinal route through the biliary tract. Hence, calcium absorption by this method may underestimate absorption of dietary calcium
but is useful for comparative purposes. It has been estimated
that bile salts may contribute about 100 g calcium/day to the
intestinal calcium contents. Bile salt calcium has been found to
be more efficiently absorbed through the intestinal mucosa than is
dietary calcium (20) but less so by other investigators (21).
     Wheat bran has been the fiber source most commonly used to
study effects of dietary fiber on calcium absorption in controlled
laboratory studies. However, wheat bran and other forms of fiber
as they occur in food products present several disadvantages in
terms of definition and by concurrently altering intakes of other
substances or materials known or suspected of having an adverse
effect on the bioavailability of calcium such as phytates and
oxalates (5,13,17,22-28). Several studies have been conducted which
have sought to separate or compare the effects of phytate and fiber

in brans (29-31). Results of McCance and Widdowson (29) indicated
that feeding dephytated whole wheat bread resulted in improved
calcium bioavailability; however, calcium bioavailability remained
inferior to that when white flour bread was fed. Reinhold et al.
(31) found that feeding purified phytate had little effect on calcium
bioavailability but that phytate-rich whole wheat bread had a pronounced negative effect. Since calcium is thought to possibly be
absorbable in the lower intestine, it may be that phytate is
digested sufficiently early to allow for the release of calcium and
its subsequent absorption but that fiber inhibits the digestion of
phytate itself thus preventing the release of calcium. Fecal
phytate excretion itself has been found to be increased with feeding of fiber-rich diets in comparison to fiber-poor diets (32).
Thus, it may be that the combination of high phytate and high fiber
diets is more detrimental to calcium absorption than is either
calcium absorption inhibitor by itself.

The term edible or dietary fiber includes a large variety of
materials with widely different physical and chemical characteristics (33-36). The only common characteristic of these substances
marking them as fiber is a resistance to digestion in the upper
gastrointestinal tract. Biologically, plant fibers may be
classified as a) structural components of cell walls (cellulose,
lignin, pectins and many hemicelluloses); b) gums and mucilages;
and c) indigestible storage polysaccharides such as guar gum.
Analytically, these fibers can be classified as water soluble
fibers (pectin, some hemicelluloses, mucilages, algal polysaccharides and storage polysaccharides) and water insoluble fibers
(cellulose, some hemicelluloses and lignin). Not only do these
different dietary fibers differ in chemical structure and physical
characteristics in vitro, but because of modifications due to
environmental changes such as pH, digestive process, presence of
bacterias and osmolality within the gastrointestinal tract and at
different sites within the gastrointestinal tract, in vivo
functional effects for any fiber type may differ considerably from
predictions made on the basis of in vitro observations. Most food
fiber sources contain mixtures of fiber substances which makes a
complex situation even more confusing. In addition such pre-
consumption treatment of dietary fiber sources such as degree of
purification, grind size, food items in which the test fibers are
contained, other food items concurrently consumed, and degree of
prehydration of the fiber content may all influence physiological
impact of test fiber consumption. The physiological characteristics of the individual consuming the fiber source perhaps has the
greatest influence on defining the apparent physiological responses
of feeding particular test fibers. Thus, it should not be surprising that there is some lack of agreement among laboratories
and even within the same laboratory on degree and even on direction
of physiological response of feeding specific test fiber sources to
humans.

In a series of human studies at the University of Nebraska, a
total of 285 human adults were fed several different fiber sources
during experimental periods of 5 to 7 days each. Some subjects
received more than one test fiber treatment; however, all subjects
received at least one fiber source as well as receiving the controlled basal diet during one period which included no added fiber

source. Test fiber sources were fed in amounts to provide about 20
g fiber/day and were fed at each of the three daily meals as part of
a bread product or mixed with peanut butter. A laboratory controlled measured diet composed of ordinary foods was fed each day
during all studies. Subjects made complete collections of urine
and stools throughout all studies. Feces were divided into period
lots through use of fecal dye markers. All studies provided similar
amounts of calcium. Milk was the primary calcium source.

Subjects were primarily young female adults, undergraduate
and graduate students at the University of Nebraska. All were
assumed to be in good health and received medical clearance for
participation prior to the start of each study from the University
of Nebraska Division of Student Health Services. The project was
approved by the University of Nebraska Institutional Review
Board Involving the Protection of Human Subjects.

Urine, feces and food were analyzed for calcium content by
atomic absorption spectrophotometry. Data were subjected to
statistical analysis by analysis of variance and Duncan's Multiple
Range Test.

Effects of the various test fibers in this project on fecal
and urine losses of calcium and on calcium balances are given in
Table I. Apparent calcium absorptions and calcium balances under
these conditions were calculated from this data and are also given
in Table I.

Addition of bran from hard red and soft white wheat bran,
psyllium fiber, and cellulose resulted in increased losses of
calcium in feces in comparison to losses when no fiber supplements
were used (P < 0.05). Urinary calcium losses were not significantly
affected; however, calcium balances were lowered when these four
fiber sources were added to the laboratory controlled diet (P < 0.05).

The adverse effects of feeding the two wheat brans and psyllium
fiber were sufficiently large to result in mean negative calcium
balances. Considering that these were young, healthy, active adult
subjects who would normally be expected to be in distinct positive
nitrogen balance at this level of calcium intake, the negative
balances were unexpected. Over an extended period of time, negative calcium balances could result in considerable body calcium
loss. Effects of feeding the other test fiber sources, corn bran,
rice bran, and pectin were not of sufficient magnitude and/or of
sufficient constancy among subjects to be found statistically significant.

In this study greater wet fecal weights were produced when the
cellulose, wheat bran and psyllium fibers were fed. Corn and rice
bran feeding produced equally high amounts of dry fecal residues
while fecal dry and wet weights produced with pectin feeding were
no greater than those produced when the basal diet without fiber
supplementation was fed. These results suggest that bulking capacity may have an important bearing on whether or not a specific
test fiber adversely influences calcium absorption.

Water-holding capacity of hemicelluloses (contained in wheat
brans and psyllium fiber) and celluloses may decrease mouth to
rectum transit time, increase fecal weight, and decrease intra-
luminal pressure (36). These characteristics might be expected to
interfere with calcium absorption decreasing time allowed for
intestinal absorption, by diluting the concentration of calcium and

Table I.   Calcium Utilization by Humans from Diets Varied in
Fiber Content

Diet[1]	Ca Intake mg/day	Urine Ca mg/day	Fecal Ca Loss mg/day	Fecal Ca Loss % Dietary Ca	% Apparent Ca Absorption	Ca Balance
Basal	1450	532	796^a	54.9^a	45.1^c	+122^a
Basal + rwb	1474	502	1032^c	70.0^c	30.0^a	-60^c
Basal + wwb	1471	551	950^{bc}	64.4^b	35.6^b	-30^{bc}
Basal + cb	1480	549	799^a	54.0^a	46.0^c	+132^a
Basal + rb	1492	512	827^{ab}	55.4^a	43.4^c	+153^a
Basal + pf	1450	580	899^b	62.0^b	38.0^b	-29^b
Basal + p	1450	551	752^a	51.9^a	48.1^c	+147^a
Basal + c	1450	543	873^b	60.2^b	39.8^b	+34^b

Note: Mean values for 127 subjects.   Values with different letter superscripts were significantly different from one another (P < 0.05).

[1] The basal diet was a measured, laboratory controlled diet based on ordinary foods fed in sufficient quantities to meet caloric requirements of the 127 subjects for weight maintenance and meeting or exceeding the National Research Council Daily Recommended Nutrient Intakes.

Diet Codes: rwb = red wheat bran; wwb = white wheat bran; cb = corn bran; rb = rice bran; pf = psyllium fiber; p = pectin; c = cellulose

by setting up a physical barrier to calcium absorption throughout the gastrointestinal tract.   Calcium as well as other nutrients may well become trapped within the matrix of fiber strands.

The gel formation ability of such water-soluble fibers as pectins and mucilages as guar gum may influence gastric emptying and increase mouth to cecum transit time.   These substances form viscous gels based on cross-linkages in the small intestines; thus, intestinal absorption with the feeding of these fibers might be expected to be decreased due to reducing rate of diffusion of calcium toward the intestinal mucosa as a result of entrapment of calcium in the gel structure.   Increase in mouth to cecum transit time, however, might be expected to result in increase in calcium absorption by allowing greater time for absorption to occur in the small intestines.   In addition, increased mouth to cecum transit

time has been found to decrease bile acid flux which might reduce
feedback inhibition of bile acid synthesis (36).

Bile acids and salts have been found to enhance the absorption
of both calcium and vitamin D; hence, to increase calcium absorption
both directly and indirectly (3,37). However, the ability of some
dietary fibers such as lignin and pectin to absorb conjugated and
deconjugated bile salts onto their surfaces to be excreted in the
feces (a mechanism credited to the hypocholesterolemic effect of some
dietary fibers) may result in an overall decrease in calcium
absorption from the gastrointestinal tract (7,33,38-40).

Several other properties of selected dietary fibers may
influence the bioavailability of calcium directly or indirectly.
Those fibers which have cation exchange capabilities such as acid
polysaccharides due to free carboxyl groups on the sugar residues
may bind minerals such as calcium (3,17,33,36). Loss of calcium
binding protein as a result of mucosa injury caused by the feeding
of some kinds of dietary fiber was credited as being the partial
cause of decreased calcium absorption in everted, rat gut sac
studies by Oku et al. (16).

In the studies discussed, wheat bran, cellulose, and psyllium
fiber feeding resulted in increased fecal fat losses and in lowered
blood serum cholesterol and triglyceride levels (14,15,32,41) as
well as increased fecal losses of calcium. Possible involvement of
dietary fat with high or low dietary fiber intake has not been
extensively investigated. However, that calcium is involved in
intestinal fat absorption is generally accepted (42-45).

Dietary Fat and Calcium Absorption

Interest in the possible connection between intake of fat and
absorption of calcium was generated by the concurrent massive
losses of calcium in patients with steatorrhea, fatty diarrhea (46,
47). Ordinarily, however, fat is very efficiently absorbed from
the gastrointestinal tract. Results of several studies in human
adults and children indicate little or no effect of level of
dietary fat on absorption of calcium (48-54). However, influence of
level of dietary fat on calcium absorption in rat studies has produced conflicting results (55-57).

In steatorrhea, the formation and precipitation of insoluble
calcium soaps formed from calcium and fatty acid has been generally
credited as the cause of reduced calcium and fat absorption.
However, Patton and Carey (42) have suggested that calcium soap
formation is a part of usual lipid digestion. Other research
indicates that the availability of calcium from calcium soaps
infused into rat intestines increases as fatty acid chain length
decreases ans as degree of polyunsaturatedness increases (58).
Basu and Nath (59), studied calcium absorption in young men eating
diets varied in levels of calcium. Diets were either nearly fat-
free or contained fat supplied primarily by butterfat, sesame oil,
peanut oil, coconut oil or mustard seed oil. Feeding of all test
fats except coconut oil resulted in a slight decrease in fecal
calcium, thus, an assumed increase in calcium bioavailability.
However, feeding of coconut oil (a highly saturated fat) resulted
in increased fecal and urinary losses of calcium.

Fat absorption has been found to be less efficient in infants than in adult humans (60-63). Thus, the potential for formation and precipitation of calcium soaps inhibiting the absorption of calcium would appear to be greater. Furthermore, formulation of products to best promote both fat and calcium utilization by bottle fed infants would appear to be a desirable, obtainable objective. Several studies have been conducted on calcium-fat interactions in human infants (64-70). Low synthesis of bile salts and low pancreatic lipase activity may be responsible for poorer fat utilization in infants than in adults (63,71). Fat from infant formulas may be lower than that from human milk because of the lack of a bile-stimulated lipase in the former (72). In infants, fat absorption tends to decrease with increase in fatty acid length, with lower degree of saturation, and with increase of total fat (3). Triglyceride structure may also influence fat absorption in the infant and, thus, indirectly, might also affect calcium absorption in the infant.

Fat and calcium from infant formulas high in stearate and palmitate have been found to be poorly utilized (70). Fecal excretion of palmitate and stearate and of calcium were correlated by Hanna et al. (64).

Both calcium and fat absorption in infants and in patients with biliary cirrhosis and subtotal gastrectomies have been improved by dietary substitution of medium chain triglycerides for long chain triglycerides (47,68,72,73). Medium chain triglycerides are more soluble than are long chain triglycerides and do not require bile salts for their digestion and absorption.

Data have been collected from several studies completed at the University of Nebraska to further investigate calcium-fat or calcium-fat-fiber interactions possibly influencing calcium utilization. In one study, 15 adolescent boys were fed measured, laboratory-controlled diets based on ordinary foods containing 886 mg of calcium and 12 g of dietary fiber from mixed sources per day. Caloric intake was maintained at levels necessary for slight weight gain in each subject. Dietary fat was maintained at 30% of the calories with the chief fat sources being safflower oil, soybean oil and apricot kernal oil. These oils have saturated to monounsaturated to polysaturated ratios of 7:11:80, 45:35:44 and 5:71:23, respectively. The 15 boys were divided into three groups of five boys each. Each group was assigned one of the three test fat diets for 20 days. Data presented are based on the last five days. Other aspects of this study relative to excreta processing and chemical and analytical analyses were as described in the earlier discussed study.

Mean urine and fecal losses of calcium of subjects while receiving the three test fats are shown on Table II. Apparent calcium absorptions and calcium balances were calculated from this data. Mean fecal calcium losses when safflower oil, soybean oil and apricot kernal oil were fed were 460, 500 and 520 mg/day. With a probability of only $P < 0.10$, these difference were not statistically significant although a strong trend toward increased fecal calcium losses with degree of saturatedness of the dietary fat was shown. Thus, a trend toward increase in calcium absorption with more polyunsaturated fats was illustrated. Coupled with changes in urinary nitrogen these changes were sufficiently large and consistent to result in changes in calcium balance which were significantly different. These results thus somewhat support those of Gacs and Barltrop (58) that fatty acid chain length may influence bioavailability of calcium.

Table II. Calcium Utilization from Low Fat Diets Varied in Saturated:Monounsaturated:Polyunsaturated Fatty Acid Ratios by Adolescent Boys

Diet[1]	Ca Intake	Urine Ca	Fecal Ca mg/day	% Dietary Ca	% Apparent Ca Absorption	Ca Balance
	mg/day	mg/day				
Safflower oil 7:11:80	886^a	210^a	460^a	51.9^a	48.1^a	216^a
Soybean oil 45:35:44	886^a	228^a	500^a	56.4^a	43.6^a	158^b
Apricot kernal oil 5:71:23	886^a	232^a	520^a	58.7^a	41.3^a	134^b

Note: Mean values for 15 subjects, 5 subjects per group. Values with different letter superscripts were significantly different from one another (P < 0.05).

[1]The diets were a measured, laboratory-controlled diet based on ordinary foods. Fats provided 30% of the dietary calories. Saturated to monounsaturated to polyunsaturated fatty acid ratios are listed below each test fat.

Effects of two levels of dietary fat on dietary calcium utilization were investigated in another study conducted at the University of Nebraska. A group of 10 healthy, young adult subjects who were fed a measured, laboratory controlled diet based on ordinary foods providing 43% of the calories from mixed sources of fat were individually age, sex, race, and weight matched to subjects from other studies consuming measured, laboratory controlled diets providing approximately the same amount of calcium but only 23% of the calories from various fat sources. Other aspects of these studies were basically as previously described.

As shown in Table III, mean fecal calcium losses tended to be higher when the higher fat diet was fed in comparison to results when the lower fat diet was fed. Therefore, apparent calcium absorption was higher when the low fat diet was fed. These differences were significant at only the P < 0.075 level; hence, only a trend was illustrated. In this study no attempt was made to equalize fatty acid proportionality patterns or cholesterol intake. These or other dietary or non-dietary factors may have influenced the observed apparent trends. Other studies with human adults have not demonstrated any apparent influence on level of dietary fat on calcium absorption.

Table III.   Calcium Utilization from High and Low Fat Diets by
Adolescent Humans

Diet	Ca Intake mg/day	Urine Ca mg/day	Fecal Ca mg/day	Fecal Ca % Dietary Ca	% Apparent Ca Absorption	Ca Balance
High fat: 43% calories from fat	1280	395	785	61.4	38.6	100
Low fat: 23.1% calories from fat	1280	433	690	53.9	46.1	157

Apparent calcium absorption of 25 human adult subjects while
receiving self-selected diets and a laboratory controlled diet providing 30% of the calories from fat, 20 g fiber from psyllium/day
and less than 50 mg cholesterol/day are shown on Table IV. Subjects
were selected from a far larger group for similarity in calcium
intake but combined with a high fat, high cholesterol, low fiber
self-selected diet. Fecal calcium losses were based on 7-day
experimental period composites as denoted by fecal dyes. Obviously,
this was not a very well controlled study; however, subjects when
consuming the laboratory controlled diet did appear to lose more
fecal calcium than they did while receiving the higher fat, lower
fiber, self-selected diet. Mean values were statistically different
from one another (P < 0.05).

Table IV.   Mean Calcium Utilization from Self-Selected and High
Fiber, Low Cholestrol, Low Fat Controlled Diets by Human Adults

Diet	Ca Intake mg/day	Urine Ca mg/day	Fecal Ca mg/day	Fecal Ca % Dietary Ca	% Apparent Ca Absorption	Ca Balance
Self-selected, 40% or less calories from fat; < 10 g fiber; > 500 g cholesterol	1462[a]	498[a]	762[a]	52.1[a]	47.9[a]	202[a]
Controlled, 30% calories from fat, 30 g fiber, < 50 g cholesterol	1440[a]	425[b]	995[a]	69.1[b]	30.9[b]	21[b]

Note: Mean values of 25 subjects. Values with different letter
superscripts are significantly different from one another (P<0.05).

These results apparently support those in the previous study; however, because dietary fiber was higher in the low fat diet than in the high fat diet, it is impossible to separate the effects of fat, fiber and possibly cholesterol. The alterations used in the low fat, controlled diet were essentially those recommended in the U.S. Dietary Goals/Guidelines while the moderately high fat, self-selected diets resembled usual American dietary intake patterns.

In order to separate these factors, data were drawn from several other studies which have been conducted at the University of Nebraska (Table V). Subjects were age, sex, weight, and race matched from four different studies providing 1450 mg calcium/day and less than 100 mg cholesterol/day but varied in fat and in fiber content. The four variations were a) high fat (45% of calories) and high fiber (30 g/day); b) high fat (45% of calories) and low fiber (8 g/day); c) low fat (30% of calories) and high fiber (30 g/day); and d) low fat (30% of calories) and low fiber (8 g/day). Ten subjects were selected for each group. Skim milk was the primary source of calcium in all diets. Ground psyllium husk was the principle fiber source in high fat diets. Hydrogenated vegetable oils were added to high fat diets at the expense of sucrose and starch. Using this approach, fiber from psyllium clearly had an adverse effect on calcium at both high and low fat intake levels. The effect of fat intake, if any, was of a much lesser degree.

Table V.   Calcium Utilization by Humans from Diets Varied in Fat and Fiber Content

Diet[1]	Ca Intake	Urine Ca	Fecal Ca[2]		% Apparent Ca Absorption	Ca Balance
			mg/day	% Dietary Ca		
	mg/day	mg/day				
High fat, high fiber	1450^a	456^a	1076^b	74.2^b	25.8^a	-82
Low fat, high fiber	1450^a	471^a	999^b	68.9^b	31.1^a	-20
High fat, low fiber	1452^a	501^b	787^a	54.2^a	45.8^b	$+164$
Low fat, low fiber	1449^a	489^b	752^a	51.9^a	48.1^b	$+208$

[1]Diet Code: High fat = 45% calories from fat; Low fat = 30% calories from fat; Low fiber = 8 g/day; High fiber = 30 g/day

[2]Mean values of 10 subjects/diet group. Values with different letter superscripts are significantly different from one another ($P < 0.05$).

Conclusion

In conclusion some but not all sources of dietary fiber have been found to have an adverse effect on the absorption of dietary fiber.

The mechanisms by which various forms of dietary fiber influence calcium bioavailability apparently also differ. In some cases, apparent dietary fiber effects on calcium bioavailability may be secondary to effects on bile acid and salt secretion and reabsorption or to other dietary components.

Factors negatively affecting absorption of dietary fat may also be expected to negatively affect absorption of calcium. In the normal, healthy adult, dietary fat absorption is very efficient. However, such factors as level of dietary fat, fatty acid chain length, degree of fatty acid saturatedness, or triglyceride structure may have a measurable effect on calcium absorption under some circumstances.

Acknowledgments

Supported by Nebraska Agricultural Experiment Station Project 91-024 and by USDA, CSRS Regional Research Projects W-143 and NC-167. Published as Nebraska Agriculture Experiment Station Journal Series Number 7621.

Literature Cited

1. "Nutrition and Your Health: Dietary Guidelines for Americans," U.S. Department of Agriculture, 1980.
2. Allen, L.H. Nutrition News 1984, 47(1), 1-3.
3. Allen, L. Am. J. Clin. Nutr. 1982, 35, 783-808.
4. McCance, R.A. and Widdowson, E.M. J. Physiol. 1942, 101, 44-85.
5. Burton, H.B. J. Biol. Chem. 1930, 85, 405-419.
6. Branch, W.J., Southgate, D.A.T., James, W.P.T. Proc. Nutr. Soc. 1975, 34, 120A.
7. Cummings, J.H., Hill, M.J., Houston, H., Branch, W.J. and Jenkins, D.J.A. Am. J. Clin. Nutr. 1979, 32, 2086-2093.
8. Cummings, J.H., Southgate, D.A.T., Branch, W.J., Wiggins, H.S., Houston, H. and Jenkins, D.J.A. Br. J. Nutr. 1979, 41, 477-485.
9. Harmouth-Hoene, A.S. and Schelenz, R. J. Nutr. 1980, 110, 1774-1784.
10. Heaton, K.W. and Pomare, E.W. Lancet, 1974, i, 49-50.
11. Ismail-Bugi, F., Reinhold, J.G., Faraji, B., and Abadi, P. J. Nutr. 1977, 107, 510.
12. James, W.P.T., Branch, W.J. and Southgate, D.A.T. Lancet, 1978, i, 638-639.
13. Kelsay, J.L., Behall, M.K. and Prather, E.S. Am. J. Clin. Nutr. 1979, 32, 1876-1880.
14. Kies, C. In: "Unconventional Sources of Dietary Fiber"; Furda, I., Ed.; ACS SYMPOSIUM SERIES, American Chemical Society: Washington, D.C., 1983; pp. 61-70.
15. McHale, M., Kies, C. and Fox, H.M. J. Food Sci. 1979, 44, 1412-1417.
16. Oku, T., Konishi, F. and Hosoya, N. J. Nutr. 1982, 112, 410-415.
17. Reinhold, J.G., Faradi, B., Parichehr, A. and Ismail-Beigi, F. J. Nutr. 1976, 106, 493-503.

18. Slavin, J.L. and Marlett, J.A. Am. J. Clin. Nutr. 1980, 33, 1932-1939.
19. Jenkins, D.J.A., Hill, M.S. and Cummings, J.E. Am. J. Clin. Nutr. 1975, 28, 1408.
20. Heaney, R.B. and Skillman, T.G. J. Lab. Clin. Med. 1964, 64, 29-41.
21. Rose, G.A., Reed, G.W. and Smith, A.H. Br. Med. J. 1965, 1, 690-692.
22. Bonner, P., Hummel, F.C., Bates, M.F., Horton, J., Hunscher, H.A., and Macy, I.G. J. Pediatr. 1938, 12, 188-196.
23. Widdowson, E.M. Nutr. Dieta. 1970, 15, 38.
24. Ford, J.A., Colhoun, E.M., McIntosh, W.B., and Dunnigan, M.G. Br. Med. J. 1972, 3, 446-447.
25. Bronner, F., Harris, R.S., Maletskos, C.J., and Benda, C.E. J. Nutr. 1954, 54, 523-542.
26. Bricker, M.L., Smith, J.M., Hamilton, T.S. and Mitchell, H.H. J. Nutr. 1949, 39, 445-461.
27. Cullumbine, H., Basnayake, V., Lomottee, J., Wickramanayake. Br. J. Nutr. 1950, 4, 101-111.
28. Schroeder, L.J., Cahill, W.M., Smith, A.H. J. Nutr. 1946, 32, 413-422.
29. McCance, R.A. and Widdowson, E.M. J. Physiol. 1942, 101, 304-313.
30. Reinhold, J.G., Ismail-Beigi, F. and Faradji, B. Nutr. Reports Int. 1973, 12, 75-85.
31. Reinhold, J.G., Khosrow, N., Lahimgarzadeh, A. and Hedayate, H. Lancet 1973, i, 283-288.
32. Kies, C., Peterson, T., Ladlie, F. and Fox, H.M. Cereal Foods World 1980, 25, 528.
33. Anderson, J.W. and Chen, W.J.L. Am. J. Clin. Nutr. 1979, 32, 346-363.
34. Kelsay, J.L. Am. J. Clin . Nutr. 1978, 31, 142-159.
35. Vanhouny, G.V. Fed. Proc. 1982, 41, 2801-2806.
36. Eastwood, M.A. and Kay, R.M. Am. J. Clin. Nutr. 1979, 32, 364-367.
37. Webling, D.D.A. and Holdsworth, E.S. Biochem. J. 1966, 100, 652-660.
38. Eastwood, M.A. and Hamilton, D. Biochem. Biophys. Acta. 1968, 1952, 167.
39. Eastwood, M.A. and Mitchell, W.D. In "Fiber in Human Nutrition" Spiller, G.A. and Amen, R., Eds.; Academic: New York, 1973; p. 303.
40. Birkner, H.J. and Kerns, F. Gastroenterology 1974, 67, 237.
41. Johnson, J., Kies, C. and Fox, H.M. Cereal Chem. (submitted), 1984.
42. Patton, J.S. and Carey, M.C. Science 1979, 204, 145-148.
43. Day, G.M., Chance, G.W., Radde, I.C., Reilly, B.J., Park, E., Sheepers, J. Pediatr. Res. 1975, 9, 568-575.
44. Lutwak, L., Laster, L., Gitelman, H.J., Fox, M. and Whedon, G.D. Am. J. Clin. Nutr. 1964, 14, 76-82.
45. Drenick, E.J. Gastroenterology 1961, 41, 242-244.
46. Aub, J.C., Tibbetts, D.M. and McLean, R. J. Nutr. 1937, 13, 635-655.
47. Agnew, J.E. and Holdsworth, C.D. Gut 1971, 12, 973-977.

48. Meyer, F.L., Brown, M.L., Wright, H.J. and Hathaway, M.L. "A Standardized Diet for Metabolic Studies. Its Development and Application." U.S. Dept. of Agriculture Technical Bull. 1126; 1955.
49. Fuqua, M.E. and Patton, M.B. *J. Amer. Dietet. Assoc.* 1953, 29, 1010–1013.
50. Steggerda, F.R., Mitchell, H.H. *J. Nutr.* 1951, 45, 201–211.
51. Holt, L.E., Courtney, A.M. and Fales, H.L. *Amer. J. Dis. Child.* 1920, 19, 97–113.
52. Holt, L.E., Courtney, A.M. and Fales, H.L. *Amer. J. Dis. Child.* 1920, 19, 201–222.
53. Holt, L.E. and Fales, H.L. *Amer. J. Dis. Child.* 1923, 25, 247–256.
54. Mallon, M.G., Jordon, R. and Johnson, M. *J. Biol. Chem.* 1930, 88, 163–167.
55. French, C.E. *J. Nutr.* 1942, 23, 375–384.
56. French, C.E. and Elliot, R.F. *J. Nutr.* 1943, 25, 17–21.
57. Speckman, E.W. and Brink, M.F. *J. Amer. Dietet. Assoc.* 1967, 51, 517–522.
58. Gacs, G., Barltrop, D. *Gut* 1977, 18, 64–68.
59. Basu, K.P. and Nath, H.P. *Indian J. Med. Res.* 1946, 34, 27–30.
60. Filer, L.J., Jr., Mattson, F.H. and Fomon, S.J. *J. Nutr.* 1970, 99, 293–298.
61. Fomon, S.J., Ziegler, E.E., Thomas, L.M., Jensen, R.L., Filer, L.J. *Am. J. Clin. Nutr.* 1979, 23, 1299–1313.
62. Katz, L.K., Hamilton, J.R. *J. Pediatr.* 1974, 85, 608–614.
63. Watkins, J.B. *Pediatr. Clin. North Amer.* 1974, 21, 502–514.
64. Hanna, F.M., Navarrette, D.A., Hsu, F.A. *Pediatrics* 1970, 45, 216–224.
65. Muhl, G. *Acta Paediat.* 1926, 5, 188–222.
66. Shaw, J.C.L. *Pediatrics* 1976, 57, 16–25.
67. Southgate, D.A.T., Widdowson, E.M., Smits, B.J., Cooke, W.T., Walker, C.H.M., Mathers, N.P. *Lancet* 1969, 1, 487–489.
68. Tanibhedhyangkul, P., Hashim, S.A. *Pediatrics* 1978, 61, 537–545.
69. Widdowson, E.M. *Lancet* 1965, 2, 1099–1105.
70. Williams, M.L., Rose, C., Morrow, G., III, Sloan, S., Barness, L. *Am. J. Clin. Nutr.* 1970, 23, 1322–1330.
71. Norman, A., Strandvik, B., Ojamae, O. *Acta Pediatr. Scand.* 1972, 61, 571–576.
72. Hamosh, M. In "Textbook of Gastroenterology and Nutrition in Infancy"; Lebenthal, E., Ed.; Raven Press: New York, 1981; pp. 473–482.
73. Kehayoglou, A.K., Williams, H.S., Whimster, W.F., Holdsworth, C.D. *Gut* 1968, 9, 597–603.

RECEIVED December 26, 1984

Author Index

Subject Index

189

Production by Meg Marshall
Indexing by Susan Robinson
Jacket design by Pamela Lewis

Elements typeset by Hot Type Ltd., Washington, D.C.
Printed and bound by Maple Press Co., York, Pa.